普通高等教育"十一五"
国家级规划教材

21世纪高等学校计算机类专业
核心课程系列教材

计算机英语实用教程

（第4版）

◎ 张强华 司爱侠 编著

清华大学出版社
北京

内 容 简 介

本书软件、硬件和网络并重，兼顾人工智能、大数据、虚拟现实等发展热点。本书分为 12 个单元，每个单元实体部分包括两篇课文、单词、词组、缩略语、课文翻译及习题；线上资源包括课文听力材料、语法手册、在线阅读材料及总词汇表。线上资源均可扫码下载。

本书为普通高等教育"十一五"国家级规划教材，既可作为高等院校信息类（包括计算机科学与工程、计算机应用与维护、计算机网络、软件工程和信息管理等专业）专业英语教材，也可作为培训班教材和从业人员的自学参考书。

图书在版编目（CIP）数据

计算机英语实用教程/张强华，司爱侠编著. —4 版. —北京: 清华大学出版社, 2021.5
21 世纪高等学校计算机类专业核心课程系列教材
ISBN 978-7-302-56972-5

Ⅰ. ①计… Ⅱ. ①张… ②司… Ⅲ. ①电子计算机—英语—高等学校—教材 Ⅳ. ①TP3

中国版本图书馆 CIP 数据核字(2020)第 231837 号

策划编辑：魏江江
责任编辑：王冰飞
封面设计：刘　键
责任校对：焦丽丽
责任印制：杨　艳
出版发行：清华大学出版社
网　　址：http://www.tup.com.cn，http://www.wqbook.com
地　　址：北京清华大学学研大厦 A 座　　邮　　编：100084
社 总 机：010-62770175　　邮　　购：010-83470235
投稿与读者服务：010-62776969，c-service@tup.tsinghua.edu.cn
质 量 反 馈：010-62772015，zhiliang@tup.tsinghua.edu.cn
课 件 下 载：http://www.tup.com.cn，010-83470236
印 装 者：三河市吉祥印务有限公司
经　　销：全国新华书店
开　　本：185mm×260mm　　印　张：17.75　　字　　数：434 千字
版　　次：2004 年 8 月第 1 版　　2021 年 7 月第 4 版　　印　　次：2021 年 7 月第 1 次印刷
印　　数：50501~52000
定　　价：49.80 元

产品编号：087751-01

前 言

计算机行业的从业人员必须快速掌握最新技术，这有赖于其专业英语能力。英语水平已经成为决定工作能力的因素之一。要提高专业英语水平，就必须进行针对性的学习。本书旨在切实提高读者实际使用计算机英语的能力。

本书出版后，深受读者喜爱，被许多学校选为教材，也有众多的从业人士自学使用。经出版社筛选推荐、教育部聘请专家评审、网络公示等程序，本书被批准为普通高等教育“十一五”国家级规划教材。在此，对各位读者的厚爱和专家的支持表示衷心感谢！

为了使本书进一步优化完善，紧跟技术发展，我们依照立体化教材理念对第 3 版进行了重大修订，使得本书由线下实体教材和线上资源两部分构成。

实体教材以 Unit 为单位，每一单元由以下几部分组成：课文——两篇课文，立足实用，软件、硬件和网络并重，同时兼顾人工智能、大数据、虚拟现实等发展热点；单词——给出课文中出现的新词，读者由此可以积累计算机专业的基本词汇；词组——给出课文中的常用词组；缩略语——给出课文中出现的、业内人士必须掌握的缩略语；参考译文——给出课文的译文，供读者学习时对照和参考；习题——包括课文练习、单词练习、短文翻译和短文填空。

线上资源包括：课文听力资源——本书配套课文听力资源（扫描每一单元两篇课文标题旁边的二维码即可播放）；语法手册——每单元一个核心语法点，系统地讲述计算机领域中常见的语法，这些内容整合起来可以作为“计算机英语简明语法手册”，有助于读者对语法进行系统学习，为读者的阅读和写作提供有力的支持（扫描每一单元末尾 Online Resources 中的二维码即可下载）；在线阅读材料——本书提供了 24 篇阅读材料，内容丰富，这些资料增加了本书的深度和广度，不仅可以进一步扩大读者的视野，也可供教师选作“翻转课堂”等教学改革的资料（扫描每一单元末尾 Online Resources 中的二维码即可下载）；总词汇表——便于读者复习、记忆单词，也可作为小词典供长期查阅（扫描前言下方的二维码即可下载）。

本书提供教学大纲、教学课件、参考试卷、参考答案等配套资源，扫描封底的“课件下载”二维码，在公众号“书圈”下载。

本书既可作为高等院校信息类（包括计算机科学与工程、计算机应用与维护、计算机网络、软件工程和信息管理等专业）专业英语教材，也可作为培训班教材和从业人员的自学参考书。

编 者

2021年3月

总词汇表

目 录

Unit 1

Computer Hardware

Text A

Computer Hardware (1)

扫码听课文

1. Introduction

Hardware is the most visible part of any information system: the equipment such as computers, scanners and printers that is used to capture data, transform it and present it to the user as output. Although we will focus mainly on the personal computer (PC) and the peripheral devices that are commonly used with it, the same principles apply to the complete range of computers:

- Supercomputers, a term used to denote the fastest computing engines available at any given time, which are used for running exceptionally demanding scientific applications.
- Mainframe computers, which provide high-capacity processing and data storage facilities to hundreds or even thousands of users operating from terminals.
- Servers, which have large data storage capacities enabling users to share files and application software, although processing will typically occur on the user's own machine.
- Workstations, which provide high-level performance for individual users in computationally intensive fields such as engineering.
- Personal computers (including laptop/notebook computers), which have a connected monitor, keyboard and CPU, and have developed into a convenient and flexible business tool capable of operating independently or as part of an organizational network.
- Mobile devices such as personal digital assistants or the latest generation of cellular telephones, which offer maximum portability plus wireless connection to the internet, although they do not offer the full functionality of a PC.

2. Input Devices

Data may enter an information system in a variety of different ways, and the input device that is the most appropriate will usually depend on the type of data being entered into the system, how frequently this is done, and who is responsible for the activity. For example, it would be more efficient to scan a page of typed text into an information system rather than retyping it, but if this happens very seldom, and if typing staff are readily available, then the cost of the scanner might not be justified. However, all of the input devices described in this chapter have at least one thing in common: the ability to translate non-digital data types such as text, sound or graphics into digital format for processing by a computer.

2.1 The Keyboard

A lot of input still happens by means of a keyboard. Usually, the information that is entered by means of a keyboard is displayed on the monitor. The layout of most keyboards is similar to that of the original typewriter on which it was modeled.

2.2 Pointing Devices

The now ubiquitous electronic mouse is an essential input device for use with any graphical user interface. Buttons on the mouse can be used to select icons or menu items, or the cursor can be used to trace drawings on the screen.

Touchscreens are computer monitors that incorporate sensors on the screen panel itself or its sides. The user can indicate or select an area or location on the screen by pressing a finger onto the monitor. Light and touch pens work on a similar principle, except that a stylus is used, allowing for much finer control. Touch pens are more commonly used with handheld computers such as personal organizers or digital assistants. They have a pen-based interface whereby a stylus is used on the small touch-sensitive screen of the handheld computer, mainly by means of ticking off pre-defined options, although the fancier models support data entry either by means of a stylized alphabet, which resembles a type of shorthand, or some other more sophisticated handwriting recognition interface.

Digitizer tablets, also known as graphics tablets, use a pressure sensitive area with a stylus. This can be used to trace drawings.

Data glove looks like a hand glove but it contains a large number of sensors and has a data cable attached, though data cable is being replaced by means of infrared cordless data transmission. Not only does the data glove allow for full three-dimensional movement but it also senses the position of individual fingers and translate this into a grip. The glove is currently used in virtual reality simulators where the user moves around in an artificially rendered environment projected onto tiny LCD screens fitted into vision goggles. The computer generates various imaginary objects, which the user can "pick up" and manipulate by means of the glove. Advanced models even allow for tactile feedback by means of small pressure pockets built into the glove.

2.3 Optical Scanners and Readers

There are a number of different optical scanner technologies on the market.

- Optical scanners use light-emitting devices to illuminate the printing on paper. Depending on how much light is reflected, a light sensor determines the position and darkness (or color) of the markings on the paper. Special-purpose optical scanners are in use by postal services to read and interpret hand-written postal codes. General-purpose scanners are used with personal computers to scan in images or text. A common use of optical scanners is the scanning of black-and-white or color images and pictures. When scanning text, it is necessary to load additional optical character recognition (OCR) software that converts the scanned raster-image of the text into the equivalent character symbols, so that they can be edited using word processing software.

- Barcode scanners detect sequences of vertical lines of different widths. These scanners have become very popular with retailers due to the fact that all pre-packaged products are now required to have a product bar code on their packaging. Libraries now also commonly use barcode scanners. They are more generally used for tracking large numbers of physical items, such as luggage handling by airlines.
- Optical mark readers are capable of reading dark marks on specially designed forms. The red multiple choice answer sheets in use at many educational and testing institutions are a good example.

2.4 Other Input Devices

A magnetic card reader reads the magnetized stripe on the back of plastic credit-card size cards. These cards need to be pre-recorded following certain standards. Although the cards can hold only a tiny amount of information, they are very popular for access control.

Biometric devices are used to verify personal identity based on fingerprints, iris or retinal scanning, hand geometry, facial characteristics etc. A scanning device is used to capture key measurements and compare them against a database of previously stored information. This type of authentication is becoming increasingly important in the control of physical access.

Finally, voice input devices are coming of age. Voicerecognition has recently made a strong entry into the market with the availability of low-cost systems that work surprisingly well with today's personal computers. These systems allow for voice control of most standard applications.

3. Central Processing Unit (CPU)

Once data has been entered into a computer, it is acted on by the CPU, which is the real brain of the computer.

3.1 Components of the CPU

- The CPU has two major components.
- The arithmetic logic unit (ALU) executes the actual instructions. It knows how to add or multiply numbers, compare data, or convert data into different internal formats.
- The control unit does the "housekeeping". It ensures that the instructions are processed on time, in the proper sequence, and operate on the correct data.

3.2 Speed of Processing

One can measure the speed of the CPU by checking the time it takes to process one single instruction. However, instead of indicating the time it takes to execute a single instruction, the processing speed is usually indicated by how many instructions (or computations) a CPU can execute in a second.

In practice, the speed of a processor is dictated by four different elements: the “clock speed”, which indicates how many simple instructions can be executed per second; the word length, which is the number of bits that can be processed by the CPU at any one time; the bus width, which determines the number of bits that can be moved simultaneously in or out of the CPU; and then the physical design of the chip, in terms of the layout of its individual transistors.

4. Main Memory

The function of main memory (also referred to as primary memory, main storage or internal storage) is to provide temporary storage for instructions and data during the execution of a program. Main memory is usually known as RAM, which stands for random access memory.

4.1 Random Access Memory (RAM)

RAM consists of standard circuit-inscribed silicon microchips that contain many millions of tiny transistors. Very much like the CPU chips, their technology follows to the law of Moore, which states that they double in capacity or power (for the same price) every 18 months. A RAM chip easily holds hundreds of megabytes (million characters). They are frequently pre-soldered in sets on tiny memory circuit boards called SIMMs (single in-line memory modules) or DIMMs (dual in-line memory modules) which slot directly onto the motherboard: the main circuit board that holds the CPU and other essential electronic elements. The biggest disadvantage of RAM is that its contents are lost whenever the power is switched off.

Two important types of RAM are:

- Cache memory is ultra-fast memory that operates at the speed of the CPU. Access to normal RAM is usually slower than the actual operating speed of the CPU. To avoid slowing the CPU down, computers usually incorporate some more expensive, faster cache RAM that sits in between the CPU and RAM. This cache holds the data and programs that are needed immediately by the CPU. Although today’s CPUs already incorporate an amount of cache on the circuit itself, this on-chip cache is usually supplemented by an additional, larger cache on the motherboard.
- Flash RAM or flash memory consists of special RAM chips. It fits into custom ports on many notebooks, hand-held computers and digital cameras. Unlike normal RAM, flash memory is nonvolatile. It holds its contents even without external power, so it is also useful as a secondary storage device.

4.2 Read-Only Memory (ROM)

A small but essential element of any computer, ROM also consists of electronic memory microchips but, unlike RAM, it does not lose its contents when the power is switched off. Its function is also very different from that of RAM. Since it is difficult or impossible to change the contents of ROM, it is typically used to hold program instructions that are unlikely to change during the lifetime of the computer. The main application of ROM is to store the so-called boot program. ROM chips are also found in many devices which contain programs that are unlikely to change over a significant period of time. Just like RAM, ROM comes in a number of different forms:

- PROM (programmable read-only memory) is initially empty and can be custom-programmed once only using special equipment. Loading or programming the contents of ROM is called burning the chip since it is the electronic equivalent of blowing tiny transistor fuses within the chip. Once programmed, ordinary PROMs cannot be modified afterwards.
- EPROM (erasable programmable read-only memory) is like PROM, but by using special equipment such as an ultraviolet light gun, the memory contents can be erased so that the EPROM can be re-programmed.
- EEPROM (electrically erasable programmable read-only memory) is similar to EPROM, but it can be re-programmed using special electronic pulses rather than ultraviolet light so no special equipment is required.

New Words

hardware	[ˈhɑ:dweə]	*n.*计算机硬件
equipment	[ɪˈkwɪpmənt]	*n.*设备，装备；器材
scanner	['skænə]	*n.*扫描设备；扫描器
printer	['prɪntər]	*n.*打印机
capture	[ˈkæptʃə]	*vt.&n.*捕获，捕捉
transform	[trænsˈfɔ:m]	*v.*转换，变换
output	[ˈaʊtpʊt]	*n.*&vt.输出
supercomputer	[ˈsu:pəkəmpju:tə]	*n.*超级计算机，巨型计算机
engine	[ˈendʒɪn]	*n.*发动机，引擎
application	[ˌæplɪˈkeɪʃn]	*n.*适用，应用，运用
mainframe	[ˈmeɪnfreɪm]	*n.*主机
process	[ˈprəʊses]	*n.*过程 *vt.*加工；处理
terminal	[ˈtɜ:mɪnl]	*adj.*终端的，末端的 *n.*终端
server	[ˈsɜ:və]	*n.*服务器
share	[ʃeə]	*v.*共享，分享
file	[faɪl]	*n.*文件
workstation	[ˈwɜ:psteɪʃn]	*n.*工作站

performance	[pəˈfɔ:məns]	*n.*表现；执行
individual	[ˌɪndɪˈvɪdʒʊəl]	*adj.*个人的，独特的，个别的
computational	[ˌkɒmpjʊˈteɪʃənl]	*adj.*计算的
laptop	[ˈlæptɒp]	*n.*便携式计算机
connect	[kəˈnekt]	*vt.*连接，联结；使……有联系
		*vi.*连接；建立关系
monitor	[ˈmɒnɪtə]	*n.*显示器；监测仪
		*vt.*监控
keyboard	[ˈki:bɔ:d]	n.键盘
independently	[ˌɪndɪ'pendəntlɪ]	*adv.*独立地，无关地
portability	[ˌpɔ:tə'bɪlɪtɪ]	*n.*可携带，轻便
wireless	[ˈwaɪələs]	*adj.*无线的
connection	[kəˈnekʃn]	*n.*连接；联系，关系；连接点
internet	[ˈɪntənet]	*n.*互联网
functionality	[ˌfʌŋkʃəˈnælɪtɪ]	*n.*功能，功能性
input	[ˈɪnpʊt]	*n.*输入；输入的数据
		*vt.*把……输入计算机
device	[dɪˈvɑɪs]	*n.*装置，设备；方法；策略；手段
text	[tekst]	*n.*文本
digital	[ˈdɪdʒɪtl]	*adj.*数字的，数据的
format	[ˈfɔ:mæt]	*n.*格式
		*vt.*使格式化
enter	[ˈentə]	*vt.&vi.*输入，进入
model	[ˈmɒdl]	*n.*模型，典型
		*vt.*模仿
ubiquitous	[ju:ˈbɪkwɪtəs]	*adj.*无所不在的，普遍存在的
mouse	[mɑʊs]	*n.*鼠标
button	[ˈbʌtn]	*n.*按钮
icon	[ˈɑɪkɒn]	*n.*光标，图标
menu	[ˈmenju:]	*n.*菜单
touchscreen	['tʌtʃskri:n]	*n.*触摸屏
sensor	[ˈsensə]	*n.*传感器，灵敏元件
panel	[ˈpænl]	*n.*面板；控制板
fancier	[ˈfænsɪə]	*n.*发烧友，对某事物有特别爱好的人
recognition	[ˌrekəgˈnɪʃn]	*n.*认识，识别

sensitive	[ˈsensətɪv]	*adj.*敏感的，灵敏的
infrared	[ˌɪnfrəˈred]	*adj.*红外线的
		*n.*红外线
cordless	[ˈkɔ:dlɪs]	*adj.*不用电线与电源相连的，无电线的
three-dimensional	[θri:dɪˈmenʃənəl]	*adj.*三维的，立体的
sense	[sens]	*n.*感觉；识别力
		*vt.*感觉，感知，感到；理解，领会
simulator	[ˈsɪmjʊleɪtə]	*n.*模拟装置，模拟器
imaginary	[ɪˈmædʒɪnərɪ]	*adj.*想象中的，假想的，虚构的
manipulate	[məˈnɪpjʊleɪt]	*vt.*操作，处理
tactile	[ˈtæktaɪl]	*adj.*触觉的，触觉感知的
feedback	[ˈfi:dbæk]	*n.*反馈，反应
scan	[skæn]	*vt.*扫描
raster	['ræstə]	*n.*光栅
equivalent	[ɪˈkwɪvələnt]	*adj.*相等的，相当的，等效的
character	[ˈkærəktə]	*n.*字符
symbol	[ˈsɪmbl]	*n.*符号，记号
		*vt.*用符号代表
barcode	[bɑ:'kəʊd]	*n.*条形码
detect	[dɪˈtekt]	*vt.*检测
verify	[ˈverɪfaɪ]	*vt.*核实；证明；判定
identity	[ɑɪˈdentɪtɪ]	*n.*身份
fingerprint	[ˈfɪŋgəprɪnt]	*n.*指纹，指印
		*vt.*采指纹
iris	[ˈɑɪrɪs]	*n.*虹膜
retinal	['retɪnl]	*adj.*视网膜的
low-cost	['ləʊkɒst]	*adj.*价格便宜的，廉价的
component	[kəmˈpəʊnənt]	*n.*部件，零件
		*adj.*组成的，构成的
instruction	[ɪnˈstrʌkʃn]	*n.*指令
compare	[kəmˈpeə]	*vt.&vi.*比较，对照
convert	[kənˈvɜ:t]	*v.*转换，转变
sequence	[ˈsi:kwəns]	*n.*序列；顺序；连续
measure	[ˈmeʒə]	*n.*测量，测度；措施；程度；尺寸
		*v.*测量；估量

bit	[bɪt]	*n.*比特（二进位制信息单位）
layout	[ˈleɪaʊt]	*n.*层，布局，安排，设计
chip	[tʃɪp]	*n.*芯片
transistor	[trænˈzɪstə]	*n.*晶体管
function	[ˈfʌŋkʃn]	*n.*功能，作用；函数 *vi.*有或起作用
temporary	[ˈtemprərɪ]	*adj.*临时的，暂时的
program	[ˈprəʊgræm]	*n.*程序 *v.*给……编写程序
silicon	[ˈsɪlɪkən]	*n.*硅
microchip	[ˈmaɪkrəʊtʃɪp]	*n.*微晶片，微型集成电路片
megabyte	[ˈmegəbaɪt]	*n.*兆字节
slot	[slɒt]	*n.*插槽
cache	[kæʃ]	*n.*高速缓冲存储
supplement	[ˈsʌplɪmənt]	*vt.&n.*增补，补充
nonvolatile	['nɒn'vɒlətaɪl]	*adj.*非易失性的，不易失的
impossible	[ɪmˈpɒsəbl]	*adj.*不可能的
modify	[ˈmɒdɪfaɪ]	*v.*修改
ultraviolet	[ˌʌltrəˈvaɪələt]	*adj.*紫外线的
reprogram	[rɪ'prəʊgræm]	*v.*重新编程，改变程序
pulse	[pʌls]	*n.*脉冲

Phrases

information system	信息系统
peripheral device	外围设备，外部设备
application software	应用软件
personal digital assistant	个人数字助理
be responsible for ...	为……负责，形成……的原因
pointing device	指点设备
graphical user interface	图形用户界面
menu item	菜单项
touch-sensitive screen	触摸屏
handheld computer	手持式计算机
handwriting recognition interface	手写识别界面
pressure sensitive area	压力敏感区域
data glove	数字手套

virtual reality	虚拟现实
pick up	拿起，拾起
optical scanner	光学扫描仪
postal code	邮政编码
word processing software	字处理软件
magnetic card	磁卡
magnetized stripe	磁条
access control	访问控制，访问管理
hand geometry	手形，掌形
facial characteristic	面部特征
voice input device	语音输入设备
be entered into	被键入，被输入
clock speed	时钟速率
word length	字长
bus width	总线宽度
law of Moore	摩尔定律
circuit board	电路板
switch off	关闭，切断
digital camera	数码相机
external power	外部电源，外部供电
secondary storage device	辅助存储设备
boot program	引导程序
ultra-violet light gun	紫外光枪

Abbreviations

PC (Personal Computer)	个人计算机
CPU (Central Processing Unit)	中央处理器
LCD (Liquid Crystal Display)	液晶显示器
OCR (Optical Character Recognition)	光学字符识别
ALU (Arithmetic and Logic Unit)	算术逻辑单元
RAM (Random Access Memory)	随机存储器
SIMMS (Single In-line Memory Modules)	单列直插内存模块
DIMMS (Dual In-line Memory Modules)	双列直插内存模块
ROM (Read-Only Memory)	只读存储器
PROM (Programmable Read-Only Memory)	可编程只读存储器
EPROM (Erasable Programmable Read-Only Memory)	可擦除可编程只读存储器

EEPROM (Electrically Erasable Programmable Read-Only Memory) 电可擦除可编程只读存储器

Text A 参考译文

计算机硬件(1)

1. 引言

硬件是任何信息系统中最容易看见的部分，如计算机、扫描仪和打印机之类的设备，用于捕获数据，对数据进行转换并将其作为输出呈现给用户。尽管我们将主要关注个人计算机（PC）及其常用的外围设备，但是相同的原理也适用于所有计算机。

- 超级计算机，用来表示在任何给定时间可用的最快的计算引擎，用于运行要求极高的科学应用程序。
- 大型计算机，为成百上千的终端用户提供高性能的处理能力和大容量的数据存储功能。
- 服务器，具有大的数据存储容量，使用户可以共享文件和应用程序软件，尽管处理过程通常会在用户自己的计算机上进行。
- 工作站，可在计算密集型领域（例如工程）中为单个用户提供高性能的处理能力。
- 个人计算机（包括膝上型计算机/笔记本计算机），具有连接的显示器、键盘和 CPU，并且已发展成为一种既方便又灵活的业务工具，能够独立运行或作为组织网络的一部分运行。
- 移动设备（例如个人数字助理或最新一代的蜂窝电话）可提供最大的便携性以及与互联网的无线连接，尽管它们不能提供 PC 的全部功能。

2. 输入设备

数据可以以各种不同的方式进入信息系统，由输入到系统中的数据类型、执行的频率以及负责该活动的人员决定哪些输入设备最合适。例如，将键入的文本页面扫描到信息系统中比重新键入它的效率更高，但是，如果这种情况很少发生，并且打字人员很容易找到，那么扫描仪的成本可能就不合理了。但是，本章中描述的所有输入设备至少有一个共同点：能够将非数字数据类型（例如文本、声音或图形）转换为数字格式以供计算机处理。

2.1 键盘

键盘仍然可以进行很多输入。通常，通过键盘输入的信息会显示在显示器上。大多数键盘的布局类似于原始打字机的布局。

2.2 指点设备

现在无处不在的电子鼠标是一个必不可少的输入设备，用于任何图形用户界面。可以使用鼠标上的按钮选择图标或菜单项，或者可以使用光标来跟踪屏幕上的图形。

触摸屏是在屏幕面板本身或其侧面装有传感器的计算机显示器。用户可以通过将手指按在显示器上来指示或选择屏幕上的区域或位置。除了使用手写笔以外，光笔和触控笔的工作原理相似，可以进行更精细的控制。触控笔更常用于手持计算机，例如个人管理器或数字助理。它们具有笔控界面，主要通过勾选预定义的选项，在掌上计算机的小型触摸屏上使用手写笔，尽管更高级的模型支持使用风格化字母（这类似于一种速记）或用其他更复杂的手写识别界面来输入数据。

数字化仪平板电脑也称为图形输入板，使用带有手写笔的压敏区域。可用于跟踪图形。

数据手套看起来像手套，但它包含大量传感器并连接了数据线，尽管数据线已被红外无线数据传输所取代。数据手套不仅可以进行完整的三维运动，而且还可以感应单个手指的位置，然后将其转化为抓握感。该手套目前用于虚拟现实模拟器中。在该模拟器中，用户在人工渲染的环境中四处移动，这些环境被投影到安装在视觉护目镜中的微型 LCD 屏幕上。计算机生成各种虚拟对象，用户可以通过手套“拾取”并进行操作。高级型号的手套甚至可以通过内置于手套中的小压力袋实现触觉反馈。

2.3 光学扫描仪和阅读器

市场上有许多不同的光学扫描仪技术。

- 光学扫描仪使用发光设备照亮纸张上的打印内容。根据反射的光量，光传感器确定纸张上标记的位置和暗度（或颜色）。邮政部门正在使用专用光学扫描仪来读取和识别手写的邮政编码。通用扫描仪与个人计算机结合可以扫描图像或文本。光学扫描仪的常见用途是扫描黑白或彩色图像和图片。扫描文本时，有必要加载附带的光学字符识别（OCR）软件，该软件将扫描的文本光栅图像转换为等效的字符符号，以便可以使用文字处理软件进行编辑。
- 条形码扫描仪可检测不同宽度的垂直线序列。由于现在要求所有预包装产品的包装上都带有产品条形码，因此这些扫描仪已在零售商中变得非常受欢迎。现在，图书馆通常也使用条形码扫描仪。它们更常用于跟踪大量的物品，例如航空公司的行李处理。
- 光学标记阅读器能够读取特殊设计形式的深色标记。许多教育和考试机构使用的红色的多项选择答案纸就是一个很好的例子。

2.4 其他输入设备

磁卡读取器读取信用卡大小的塑料卡背面的磁条。这些卡需要按照某些标准预先记录。尽管这些卡只能容纳极少量的信息，但它们在访问控制中非常受欢迎。

生物识别设备用于根据指纹、虹膜或视网膜扫描、手部几何形状、面部特征等来验证个人身份。扫描设备用于捕获关键测量值并将它们与先前存储的信息的数据库进行比较。这种类型的身份验证在物理访问控制中变得越来越重要。

最后，语音输入设备已经成熟。语音识别技术最近凭借大量低成本的系统进入了市场，这些系统可与当今的个人计算机完美配合。这些系统允许对大多数标准应用程序进行语音控制。

3. 中央处理器（CPU）

数据输入到计算机后，将由 CPU 对其进行处理，CPU 是计算机的真正大脑。

3.1 CPU 组件

CPU 有两个主要组件。

- 算术逻辑单元（ALU）执行实际指令。它知道如何加或乘数字，比较数据或将数据转换为不同的内部格式。
- 控制单元执行“内部处理”。它确保按时、按正确的顺序处理指令，并对正确的数据进行操作。

3.2 处理速度

可以通过检查处理一条指令所需的时间来测量 CPU 的速度。但是，除了显示执行一条指令所花费的时间外，处理速度通常由一秒内 CPU 可以执行多少条指令（或计算）来表示。

实际上，处理器的速度由 4 个不同的要素决定：“时钟速度”，它表示每秒可以执行多少条简单指令；字长，即在任何时候 CPU 可以处理的位数；总线宽度，确定可以同时移入或移出 CPU 的位数；然后是芯片的物理设计，即各个晶体管的布局。

4. 主存

主存储器（也称为主要存储器、主存或内部存储器）的功能是在程序执行期间为指令和数据提供临时存储。主存储器通常称为 RAM，它代表随机存取存储器。

4.1 随机存取存储器（RAM）

RAM 由包含数百万个微型晶体管的标准刻写电路硅芯片组成。与 CPU 芯片非常相似，它们的技术遵循摩尔定律，该定律指出，每 18 个月它们的容量或性能就会增加一倍（以相同的价格）。一个 RAM 芯片很容易容纳数百兆字节（百万个字符）。它们通常预先焊接在称为 SIMMs（单列直插式内存模块）或 DIMMs（双列直插式内存模块）的微型存储电路板上，这些电路板直接插入主板，主板是用于固定 CPU 和其他组件的主电路板。RAM 的最大缺点是，一旦关闭电源，其内容就会丢失。

RAM 的两种重要类型是：

- 高速缓存是一种以 CPU 的速度运行的超快内存。访问普通 RAM 的速度通常比 CPU 的实际运行速度慢。为了避免降低 CPU 的速度，计算机通常会在 CPU 和 RAM 之间集成一些更昂贵、速度更快的缓存 RAM。该高速缓存保存 CPU 立即需要的数据和程序。尽管当今的 CPU 已经在电路本身上集成了一定数量的高速缓存，但通常还要在主板上增加一个更大的高速缓存以提供片载高速缓存。
- Flash RAM 或闪存由特殊的 RAM 芯片组成。它适合许多笔记本计算机、手持计算机和数码相机的自定义端口。与普通 RAM 不同，闪存是非易失性的。即使没有外部电源，它也可以保存其内容，因此它也可用作辅助存储设备。

4.2 只读存储器（ROM）

ROM 是任何计算机中一个很小但必不可少的单元，它也由电子存储微芯片组成，但是与 RAM 不同，ROM 在关闭电源时不会丢失其内容。它的功能也与 RAM 完全不同。由于很难或不可能更改 ROM 的内容，因此它通常用于保存在计算机寿命期内不太可能更改的程序指令。ROM 的主要应用是存储所谓的引导程序。ROM 芯片也出现在许多设备中，这些设备包含的程序在相当长的时间内不太可能更改。就像 RAM 一样，ROM 有多种形式：

- PROM（可编程只读存储器）最初为空，只能使用特殊设备进行一次自定义编程。加载或编程 ROM 的内容称为烧录芯片，因为这等效于在芯片内烧制微小的晶体管熔丝。一旦编程，普通的 PROM 就无法修改。
- EPROM（可擦可编程只读存储器）类似于 PROM，但是通过使用特殊设备（例如紫外光枪）可以擦除存储内容，以便可以对 EPROM 进行重新编程。
- EEPROM（电可擦可编程只读存储器）与 EPROM 相似，但是可以使用特殊的电子脉冲而不是紫外线对它进行重新编程，因此不需要特殊的设备。

Text B

Computer Hardware (2)

扫码听课文

5. Secondary Storage Devices

Since the main memory of a computer has a limited capacity, it is necessary to retain data in secondary storage between different processing cycles. This is the medium used to store the program instructions as well as the data required for future processing. Most secondary storage devices in use today are based on magnetic or optical technologies.

5.1 Disk Drives

The disk drive is the most popular secondary storage device, and is found in both mainframe and microcomputer environments. The central mechanism of the disk drive is a flat disk, coated with a magnetizable substance. As this disk rotates, information can be read from or written to it by means of a head. The head is fixed on an arm and can move across the radius of the disk. Each position of the arm corresponds to a "track" on the disk, which can be visualized as one concentric circle of magnetic data. The data on a track is read sequentially as the disk spins underneath the head. There are quite a few different types of disk drives.

In Winchester hard drives, the disk, access arm and read/write heads are combined in one single sealed module. This unit is not normally removable. Since the drives are not handled physically, they are less likely to be contaminated by dust and therefore much more reliable. Mass production and technology advances have brought dramatic improvements in the storage capacity.

Large organizations such as banks, telcos and life insurance companies require huge amounts of storage space, often in the order of many terabytes (one terabyte is one million megabytes or a trillion characters). This was typically provided by a roomful of large, high-capacity hard drive units. Currently, they are being replaced increasingly by redundant arrays of independent disks (RAIDs). A RAID consists of an independently powered cabinet that contains a number (10 to 100) of microcomputer Winchester-type drives but functions as one single secondary storage unit. The advantage of the RAID is its high-speed access and relatively low cost. In addition, a RAID provides extra data security by means of its fault-tolerant design whereby critical data is mirrored (stored twice on different drives) thus providing physical data redundancy. Should a mirrored drive fail, the other drive steps in automatically as a backup.

5.2 Optical Disk Storage

Optical disks, on the other hand, are rapidly becoming the storage medium of choice for the mass distribution of data/programs and the backup of data. Similar to disk storage, information is stored and read from a circular disk. However, instead of a magnetic read head, a tiny laser beam is used to detect microscopic pits burnt onto a plastic disk coated with reflective material. The pits determine whether most of the laser light is reflected back or scattered, thus making for a binary "on" or "off". In contrast to hard disks, data is not stored in concentric cylinders but in one long continuous spiral track.

5.3 SSD

An SSD (solid state drive) is a type of mass storage device similar to a hard disk drive (HDD). It supports reading and writing data and maintains stored data in a permanent state even without power.

Unlike hard drives, SSDs do not have any moving parts (which is why they are called solid state drives). Instead of storing data on magnetic platters, SSDs store data using flash memory. Since SSDs have no moving parts, SSDs can access data faster than HDDs.

SSDs have several other advantages over hard drives as well. For example, the read performance of a hard drive declines when data gets fragmented, or split up into multiple locations on the disk. The read performance of an SSD does not diminish no matter where the data is stored on the drive. Therefore defragmenting an SSD is not necessary. Since SSDs do not store data magnetically, they are not susceptible to data loss even if there are strong magnetic fields in close proximity to the drive. Additionally, since SSDs have no moving parts, there is far less chance of a mechanical breakdown. SSDs are also lighter, quieter, and use less power than hard drives.

However, SSDs also have some disadvantages. Since the SSD technology is much newer than traditional hard drive technology, the price of SSDs is substantially higher. Most SSD drives sold today have much smaller capacities than comparable hard drives. As the SSD technology improves and the prices continue to fall, it is likely that solid state drives will replace hard disk drives for most purposes.

6. Output Devices

The final stage of information processing involves the use of output devices to transform computer-readable data back into an information format that can be processed by humans. As with input devices, when deciding on an output device you need to consider what sort of information is to be displayed, and who is intended to receive it.

One distinction that can be drawn between output devices is that of hardcopy versus softcopy devices. Hardcopy devices (printers) produce a tangible and permanent output whereas softcopy devices (display screens) present a temporary, fleeting image.

6.1 Display Screens

The desk-based computer screen is the most popular output device. The standard monitor works on the same principle as the normal TV tube: a "ray" gun fires electrically charged particles onto a specially coated tube (hence the name cathode-ray tube or CRT). When the particles hit the coating, the "coating" is being "excited" and emits light. A strong magnetic field guides the particle stream to form the text or graphics on your familiar monitor.

A technology that has received much impetus from the fast-growing laptop and notebook market is the liquid crystal display (LCD). LCDs have matured quickly, increasing in resolution, contrast, and colour quality. Their main advantages are lower energy requirements and their thin, flat size. Although alternative technologies are already being explored in research laboratories, they currently dominate the "flat display" market.

Organic light-emitting diodes (OLED) can generate brighter and faster images than LED technology, and require thinner screens.

Another screen-related technology is the video projection unit. It was originally developed for the projection of video films. Today's units fit easily into a small suitcase and project a computer presentation in very much the same way a slide projector shows a slide presentation.

6.2 Printer

Printers are a type of computer peripheral device that fall into two broad categories: 2D printers that print text and graphics onto paper (or other media) and 3D printers that create physical objects.

6.2.1 2D Printers

2D printers are by far the most common type of printer. This category can be subdivided based on the type of technology used to transfer images onto paper. Modern printers generally fall into one of the following categories:

- Inkjet: It sprays ink at a sheet of paper. Inkjet printers produce high-quality text and graphics.
- Laser: It uses the same technology as copy machines. Laser printers produce very high-quality text and graphics.
- LED: It is similar to a laser printer but uses light-emitting diodes rather than a laser to produce an image on the drum.
- Thermal printer: It works by pushing heated pins against heat-sensitive paper. Thermal printers are widely used in ATMs and cash registers.

6.2.2 3D Printers

3D printers work by depositing layers of material on top of each other to create a physical object. This type of process is also sometimes called additive manufacturing. Currently, companies are investing in a lot of research and development around 3D printing, and the technology is changing rapidly. 3D printing is expected to grow in popularity as the technology improves and costs for 3D printers decline.

6.3 Plotters

Plotters are mainly used for engineering and architectural drawings. A plotter consists of one or several — in the case of color plotters — pens affixed to an arm. As the arm moves across the sheet of paper, the pen draws lines onto the paper. It is ideal for line drawings such as plans.

6.4 Audio Output Devices

Audio output is becoming increasingly popular. There are some different types of audio output.

- Sound output is required by most multimedia applications and sophisticated games. The sound card in many of today's personal computers synthesizes sound by drawing from a library of stored sounds, essentially using the same process as found in music keyboards. More advanced multimedia workstations are equipped for full stereo multi-channel surround sound and easily surpass many modern Hi-Fi systems in cabling and speaker complexity.
- Speech synthesis is the production of speech-like output using an artificial voice. Although the lack of intonation still makes the voice sound artificial, the technology is

reasonably mature and can be found anywhere.

New Words

retain	[rɪˈteɪn]	*vt.*保持
magnetic	[mægˈnetɪk]	*adj.*磁性的
optical	[ˈɒptɪkl]	*adj.*光学的
mechanism	[ˈmekənɪzəm]	*n.*机制；（机械）结构，机械装置
substance	[ˈsʌbstəns]	*n.*物质，材料
track	[træk]	*n.*磁道，轨道
concentric	[kənˈsentrɪk]	*adj.*同一中心的，同轴的
removable	[rɪˈmu:vəbl]	*adj.*可移动的，抽取式的
reliable	[rɪˈlaɪəbl]	*adj.*可靠的，可信赖的
improvement	[ɪmˈpru:vmənt]	*n.*改进，改善，改良，增进
telco	[ˈtelkəʊ]	*abbr.*电信公司
huge	[hju:dʒ]	*adj.*巨大的，庞大的，极大的
terabyte	[ˈterəbaɪt]	*n.*太字节
fault-tolerant	[fɔ:lt-ˈtɒlərənt]	*adj.*容错的
mirror	[ˈmɪrə]	*vt.*镜像，映射
redundancy	[rɪˈdʌndənsɪ]	*n.*冗余
backup	[ˈbækʌp]	*n.*备份文件
		*adj.*备份的，备用的
distribution	[ˌdɪstrɪˈbju:ʃn]	*n.*分配，分布
microscopic	[ˌmaɪkrəˈskɒpɪk]	*adj.*微小的，细微的
reflective	[rɪˈflektɪv]	*adj.*反射的，反光的
scatter	[ˈskætə]	*vt.*（使）散开，（使）分散
		*vi.*散开，分散
cylinder	[ˈsɪlɪndə]	*n.*圆柱
spiral	[ˈspaɪrəl]	*n.*螺旋（线）
		*v.*使成螺旋形
		*adj.*螺旋形的
permanent	[ˈpɜ:mənənt]	*adj.*永久（性）的，不变的，持久的
fragment	[ˈfrægmənt]	*n.*碎片，片段；（将文件内容）分段
		vt.（使）碎裂，破裂
susceptible	[səˈseptəbl]	*adj.*易受影响的，易受感染的
replace	[rɪˈpleɪs]	*vt.*替换，代替

display	[dɪˈspleɪ]	*n*.显示器 *v*.显示
distinction	[dɪˈstɪŋkʃn]	*n*.区别；特质
hardcopy	['hɑ:dkɒpɪ]	*n*.硬拷贝
softcopy	['sɒftkɒpɪ]	*n*.软拷贝
printer	['prɪntə]	*n*.打印机
tangible	[ˈtændʒəbl]	*adj*.确实的，实际的；有形的
particle	[ˈpɑ:tɪkl]	*n*.粒子
contrast	[ˈkɒntrɑ:st]	*n*.对比度
energy	[ˈenədʒɪ]	*n*.能量
slide	[slɑɪd]	*n*.幻灯片
projector	[prəˈdʒektə]	*n*.放映机，幻灯机；投影仪
inkjet	['ɪŋkdʒet]	*adj*.喷墨的
spray	[spreɪ]	*v*.喷，喷射
laser	[ˈleɪzə]	*n*.激光；激光器
diode	[ˈdɑɪəʊd]	*n*.二极管
deposit	[dɪˈpɒzɪt]	*v*.放下，放置；使沉积，使沉淀
plotter	[ˈplɒtə]	*n*.绘图仪，绘图机
multimedia	[ˌmʌltɪˈmi:dɪə]	*n*.多媒体 *adj*.多媒体的
synthesize	[ˈsɪnθəsaɪz]	*v*.合成；综合
stereo	[ˈsterɪəʊ]	*adj*.立体声的；有立体感的 *n*.立体声，立体声系统
multi-channel	['mʌltɪ-ˈtʃænl]	*n*.多通道
surround	[səˈrɑʊnd]	*vt*.环绕，包围
complexity	[kəmˈpleksɪtɪ]	*n*.复杂性
artificial	[ˌɑ:tɪˈfɪʃl]	*adj*.人造的，人工的，人为的
intonation	[ˌɪntəˈneɪʃn]	*n*.语调，声调
reasonably	[ˈri:znəblɪ]	*adv*.相当地，合理地

Phrases

sealed module	密封模块
be contaminated by...	被……污染
storage capacity	存储容量
storage space	存储空间

in the order of	大约
a roomful of	一屋子的
optical disk	光盘
laser beam	激光束
computer-readable data	计算机可读的数据
magnetic field	磁场
video projection unit	视频投影单元
slide presentation	幻灯片演示
inkjet printer	喷墨打印机
copy machine	复印机
laser printer	激光打印机
light-emitting diode	发光二极管
thermal printer	热敏打印机，热转印打印机
cash register	现金出纳机，收银机

Abbreviations

RAID (Redundant Arrays of Independent Disk)	独立磁盘冗余阵列
SSD (Solid State Drive)	固态硬盘
HDD (Hard Disk Drive)	硬盘驱动器
CRT (Cathode-Ray Tube)	阴极射线管
OLED (Organic Light-Emitting Diodes)	有机发光二极管
ATM (Automatic Teller Machine)	自动柜员机
2D (2 Dimensions)	二维，两个维度，两个坐标
3D (3 Dimensions)	三维，三个维度，三个坐标
Hi-Fi (High-Fidelity)	高保真

Text B 参考译文

计算机硬件（2）

5. 辅助存储设备

由于计算机的主存储器容量有限，因此有必要在不同的处理周期之间将数据保留在辅助存储器中。辅助存储设备用于存储程序指令以及将来处理所需数据的介质。当今使用的大多数辅助存储设备都是基于磁或光技术的。

5.1 磁盘驱动器

磁盘驱动器是最流行的辅助存储设备，在大型机和微型计算机环境中都可以找到。磁盘驱动器的中心机构是一块平盘，上面涂有可磁化物质。当该磁盘旋转时，可以通过磁头读取或写入信息。磁头固定在磁头臂上，可以在磁盘半径范围内移动。磁头臂的每个位置对应于磁盘上的一个“磁道”，可以将其视为一个磁数据同心圆。当磁盘在磁头下方旋转时，将顺序读取轨道上的数据。有很多不同类型的磁盘驱动器。

在温彻斯特硬盘驱动器中，磁盘、磁头臂和读/写磁头组合在一个密封模块中。这个部分通常是不能取出的。由于不能对驱动器进行物理操作，因此它们不太可能被灰尘污染，所以可靠性更高。大规模生产和技术进步带来了存储容量的巨大提高。

如银行、电信公司和人寿保险公司之类的大型组织需要大量的存储空间，容量通常大约在数太字节（1太字节为一百万兆字节或一万亿个字符）。过去这通常是由一大堆大容量的硬盘驱动器提供的。当前，它们越来越多地被独立磁盘的冗余阵列（RAID）取代。RAID由独立供电的机柜组成，该机柜包含多个（10～100 个）微计算机温彻斯特型驱动器，但充当单个辅助存储单元。RAID的优点是其访问速度高且成本较低。此外，RAID通过其容错设计提供了更高的数据安全性，对关键数据进行了镜像（两次存储在不同的驱动器上），从而提供了物理数据冗余。如果镜像驱动器发生故障，其他驱动器将自动作为备份启动。

5.2 光盘存储

另外，光盘正迅速成为数据/程序的大规模分发和数据备份的首选存储介质。与磁盘存储类似，信息是从圆形磁盘存储和读取的。但是不是用磁读取头，而是使用微小的激光束来检测烧在涂有反射材料的塑料盘上的微小凹坑。凹坑确定大多数激光是反射回来还是散射出去，从而形成二进制的“开”或“关”。与硬盘相反，数据不是存储在同心圆柱体中，而是存储在一个长的连续螺旋轨道中。

5.3 固态硬盘

SSD是类似于硬盘驱动器（HDD）的一种大容量存储设备。它支持读取和写入数据，并且即使没有电源也可以将存储的数据永久保存。

与硬盘驱动器不同，SSD 没有任何活动部件（这就是它们被称为固态驱动器的原因）。SSD不用磁盘存储数据，而是使用闪存存储数据。由于SSD没有活动部件，因此，SSD可以比HDD更快地访问数据。

与硬盘驱动器相比，SSD还具有其他一些优势。例如，当数据碎片化或拆分到磁盘上的多个位置时，硬盘驱动器的读取性能会下降。无论数据存储在驱动器上的什么位置，SSD的读取性能都不会降低。因此，无须对SSD进行碎片整理。由于SSD不用磁性存储数据，即使驱动器附近有强磁场，也不会丢失数据。此外，由于SSD没有活动部件，因此发生机械故障的机会要少得多。与硬盘驱动器相比，SSD 更加轻便、安静且省电。

然而 SSD 也有一些缺点。由于 SSD 技术比传统的硬盘技术要新得多，因此 SSD 的价格要高得多。如今出售的大多数 SSD 驱动器的容量都比同类硬盘小得多。随着 SSD 技术的改进和价格的不断下降，固态驱动器很可能会在大多数情况下取代硬盘驱动器。

6. 输出设备

信息处理的最后阶段涉及使用输出设备将计算机可读数据转换回可以由人类处理的信息格式。与输入设备一样，在确定输出设备时，需要考虑要显示什么类型的信息以及由谁来接收信息。

输出设备之间的区别是硬拷贝与软拷贝。硬拷贝设备（打印机）产生有形的、永久性的输出，而软拷贝设备（显示屏）显示暂时的、短暂的图像。

6.1 显示屏

基于桌面的计算机屏幕是最受欢迎的输出设备。标准显示器的工作原理与普通电视管相同："射线"枪将带电粒子发射到经过特殊涂层的管上（因此称为阴极射线管或 CRT）。当粒子撞击涂层时，"涂层"被"激发"并发光。强磁场引导粒子流在你熟悉的显示器上形成文本或图形。

笔记本计算机的市场的迅速扩展极大地推动了液晶显示器（LCD）技术的发展。LCD 已迅速成熟，分辨率、对比度和颜色质量得到了提高。它的主要优点是有较低的电量需求以及薄而扁平的尺寸。尽管研究实验室已经在探索替代技术，但它目前在"平板显示器"市场上占主导地位。

与 LED 技术相比，有机发光二极管（OLED）可以产生更明亮、更快的图像，并且需要更薄的屏幕。

与屏幕相关的另一项技术是视频投影单元。它最初是为视频电影的放映而开发的。如今的装置可以很容易装入一个小手提箱，并以类似于幻灯机显示幻灯演示的方式投影计算机演示。

6.2 打印机

打印机是计算机外围设备的一种，分为两大类：将文本和图形打印到纸张（或其他介质）上的 2D 打印机和创建物品的 3D 打印机。

6.2.1 2D 打印机

到目前为止，2D 打印机是最常见的打印机类型。可以根据将图像转印到纸张上的技术类型来细分此类别。现代打印机通常属于以下类别之一。

- 喷墨：在一张纸上喷射墨水。喷墨打印机可产生高质量的文本和图形。
- 激光：使用与复印机相同的技术。激光打印机可产生非常高质量的文本和图形。

- LED：与激光打印机相似，但它使用发光二极管而不是激光在感光鼓上产生图像。
- 热敏打印机：通过将加热的针推到热敏纸上来工作。热敏打印机广泛用于 ATM 和收银机。

6.2.2 3D 打印机

3D 打印机的工作原理是用材料层层沉积以创建一个物理对象。这类过程有时也称为增材制造。当前，一些公司正在围绕 3D 打印进行大量研究和开发投资，并且该技术正在迅速变化。随着技术的进步和 3D 打印机成本的下降，预计 3D 打印将越来越受欢迎。

6.3 绘图仪

绘图仪主要用于绘制工程图和建筑图。绘图仪由一根或几根（对于彩色绘图仪而言）固定在手臂上的笔组成。当手臂在纸上移动时，笔会在纸上画线。它非常适用于线条图，例如平面图。

6.4 音频输出设备

音频输出变得越来越流行。有一些不同类型的音频输出。

- 大多数多媒体应用程序和复杂游戏都需要声音输出。当今许多个人计算机中的声卡都是通过从存储的声音库中提取声音来合成声音的，基本上使用与音乐键盘相同的过程。更高级的多媒体工作站配备完整的立体声多声道环绕声，并在布线和扬声器复杂性方面轻松超过许多现代的高保真音响系统。
- 语音合成是使用人工语音产生类似语音的输出。尽管缺少语调使语音听起来很虚假，但该技术相当成熟，到处可见。

Exercises

[Ex. 1] Answer the following questions according to Text A.

1. What are supercomputers?
2. What do personal computers have? What have they developed into?
3. What will the input device that is the most appropriate usually depend on?
4. What are touchscreens? What can the user do by pressing a finger onto the monitor?
5. What does the data glove do? Where is the data glove currently used?
6. What are the different optical scanner technologies on the market?
7. What are the two major components the CPU has?
8. What is main memory also referred to as? What is the function of main memory?
9. What does RAM stand for? What are the two important types of RAM?

10. What does ROM stand for? What is the main application of ROM?

[Ex.2] Answer the following questions according to Text B.

1. What is the disk drive? What is the central mechanism of the disk drive?
2. What does a RAID consist of? What is the advantage of the RAID?
3. What is an SSD?
4. What is the most popular output device?
5. What are the main advantages of LCDs?
6. What are printers? What are the two broad categories they fall into?
7. What categories do modern printers generally fall into?
8. How do 2D printers work?
9. What are plotters mainly used for?
10. What are the different types of audio output mentioned in the passage?

[Ex. 3] Translate the following terms or phrases from English into Chinese and vice versa.

1. boot program	1.
2. graphical user interface	2.
3. peripheral device	3.
4. virtual reality	4.
5. word length	5.
6. *n.*高速缓冲存储	6.
7. *n.*芯片	7.
8. *n.*功能，功能性	8.
9. *adj.*非易失性的，不易失的	9.
10. *n.*打印机	10.

[Ex. 4] Translate the following sentences into Chinese.

Main Components of a Computer

CPU is considered the most important component in a computer and for good reason. It handles most operations, by processing instructions and giving signals out to other components. The CPU is the main bridge between all the computer's major parts.

RAM is a computer component where data used by the operating system and software applications is stored so that the CPU can process them quickly. Everything stored on RAM is lost if the computer is shut off. Depending on the applications you use, there is typically a maximum limit of RAM you will need for the computer to function properly.

HDD — Also known as hard disk drive, it is the component where photos, apps, documents and such are kept. Although they are still being used, we have much faster types of storage devices such as solid state drives (SSD) that are also more reliable.

Motherboard — There is no acronym for this component but without it, there can't be a computer. The motherboard acts as the home for all other components, allows them to communicate with each other and gives them power in order to function. There are components that don't require a physical connection to the motherboard in order to work, such as Bluetooth or Wi-Fi but, if there is no connection or signal what so ever, the computer won't know it's there.

Video and sound cards — Two components which help the user interact with the computer. Although one can use a computer with a missing sound card, it's not really possible to use it without a video card. The sound card is used mainly to play sound through a speaker. However, a video card is used to send images on the screen. Without it, it would be like looking at an empty monitor.

Network adapter — Even though it is not actually required to operate the computer, the network adapter improves the user's experience as it provides access to the internet. Modern computers with operating systems such as Windows will not offer the user all of its features without an Internet connection.

[Ex. 5] Fill in the blanks with the words given below.

components	peripherals	peripheral	memory	computer
output	external	motherboard	devices	hardware

What Are Computer Peripherals?

Computer peripherals are the devices you use to expand your system's functionality and are not essential for the computer to work. Many peripherals are ___1___ devices that give you a way to input information. For example, you might use a mouse or trackpad to navigate the screen, a keyboard to type text or a microphone to record audio. Other peripherals are ___2___ devices that let you see, print or listen to something, such as monitors, printers and speakers. While debatable since they are key ___3___ for a functional computer, internal devices like hard drives and memory may be considered peripherals.

1. Keyboard, Mouse and Touch Pad Devices

Keyboards, mice, touch pads and other human interface devices are very common ___4___ of computer systems. A HID is a peripheral that lets the computer use input data or interact with the computer. HIDs are always considered peripheral ___5___ because it is possible for a computer to operate and do countless jobs without the need for human interaction. You can disconnect an HID from a ___6___ and replace it with a new one without affecting the core functionality of the system.

2. Data Storage Devices

Data storage devices are ___7___ devices capable of holding information and include system

RAM, internal hard drives, external hard drives, solid state drives, flash drives, ___8___ cards and cassettes. Any type of storage device that connects outside of the computer, or isn't necessary for the computer to run, is considered a ___9___. For example, you can connect and disconnect external hard drives, flash drives and memory cards without disabling the the computer. However, system RAM and internal hard drives straddle the peripheral line. While the ___10___ and CPU have small amounts of built-in memory, alone, they are not enough to utilize most of the computer's capabilities.

Online Resources

二维码	内　容
	计算机专业常用语法（1）：定语从句
	在线阅读（1）：Motherboard（主板）
	在线阅读（2）：Parallel Computing（并行计算）

Unit 2

Software

Text A

Different Types of Software

扫码听课文

Software is a computer program that provides instructions and data to execute user's commands.

1. Application Software

As a user of technology, application software or "apps" are what you engage with the most. They are productive end-user programs that help you perform tasks. The following are some examples of application software that allow you to do specific work:

- MS Excel: It is a spreadsheet software that you can use for presenting and analyzing data.
- Photoshop: It is a photo editing application software by Adobe. You can use it to visually enhance, catalog and share your pictures.
- Skype: It is an online communication app that you can use for video chat, voice calling and instant messaging.

Software applications are also referred to as non-essential software. They are installed and operated on a computer based on the user's requirement. There are plenty of application software that you can use to perform different tasks. The number of such apps keeps increasing with technological advances and the evolving needs of the users. You can categorize these software types into different groups, as shown in the following Table 2-1.

Table 2-1 Application Software Type and Examples

Application Software Type	Examples
Word processing software: Tools that are used to create Word sheets and type documents etc.	Microsoft Word, WordPad, AppleWorks and Notepad
Spreadsheet software: Software used to compute quantitative data	Apple Numbers, Microsoft Excel and Quattro Pro
Database software: Used to store data and sort information	Oracle, MS Access and FileMaker Pro

Continued

Application Software Type	Examples
Application Suites: A collection of related programs sold as a package	OpenOffice, Microsoft Office
Multimedia software: Tools used for a mixture of audio, video, image and text content	Real Player, Media Player
Communication Software: Tools that connect systems and allow text, audio, and video-based communication	MS NetMeeting, IRC, ICQ
Internet Browsers: Used to access and view websites	Netscape Navigator, MS Internet Explorer, and Google Chrome
Email Programs: Software used for emailing	Microsoft Outlook, Gmail, Apple Mail

2. System Software

System software helps the user, hardware, and application software to interact and function together. This type of computer software allows an environment or platform for other software and applications to work in. This is why system software is essential in managing the whole computer system.

When you first power up your computer, it is the system software that is initially loaded into memory. Unlike application software, the system software is not used by end-users. It only runs in the background of your device at the most basic level while you use other application software. This is why system software is also called "low-level software".

Operating systems are an example of system software. All of your computer-like devices run on an operating system, including your desktop, laptop, smartphone, and tablet, etc. Here is a list of examples of an operating system. Let's take a look and you might spot some familiar names of system software:

- For desktop computers, laptops and tablets: Microsoft Windows, Mac (for Apple devices), Linux.
- Other than operating systems, some people also classify programming software and driver software as types of system software.

3. Programming Software

Programming software are programs that are used to write, develop, test, and debug other software, including apps and system software. For someone who works at a bespoke software development company, for example, this type of software would make their life easier and more efficient.

Programming software is used by software programmers as translator programs. They are facilitator software used to translate programming languages (i.e., Java, C++, Python, PHP, BASIC, etc) into machine language code. Translators can be compilers, interpreters and assemblers. You can understand compilers as programs that translate the whole source code into machine code and execute it. Interpreters run the source code as the program is run line by line. And assemblers translate the basic computer instructions — assembly code — into machine code.

Different programming language editors, debuggers, compilers and integrated development environment (IDE) are an example of programming software.

4. Driver Software

Driver software is often classified as one of the types of system software. They operate and control devices and peripherals plugged into a computer. Drivers are important because they enable the devices to perform their designated tasks. They do this by translating commands of an operating system for the hardware or devices, assigning duties. Therefore, each device connected with your computer requires at least one device driver to function.

Since there are thousands of types of devices, drivers make the job of your system software easier by allowing it to communicate through a standardized language. Some examples of driver software that you may be familiar with are: printer driver, mouse driver.

Usually, the operating system comes built-in with drivers for mouse, keyboard, and printers by default. They often do not require third-party installations. But for some advanced devices, you may need to install the driver externally. Moreover, if you use multiple operating systems like Linux, Windows and Mac, then each of these supports different variants of drivers. For them, separate drivers need to be maintained for each.

5. Another Classification of Software

Let's discuss five additional subcategories of software and understand them using examples of trendy software.

5.1 Freeware

Freeware software is any software that is available to use for free. They can be downloaded and installed over the internet without any cost. Some well-known examples of freeware are: Google Chrome, Skype, Instagram, Snapchat, Adobe reader.

Although they all fall under the category of application or end-user software, they can further be categorized as freeware because they are free for you to use.

5.2 Shareware

Shareware, on the other hand, are software applications that are paid programs, but are made available for free for a limited period of time known as "trial period". You can use the software without any charges for the trial period but you will be asked to purchase it for use after the trial ends. Shareware allows you to test drive the software before you actually invest in purchasing it. Some examples of shareware that you must be familiar with are: Adobe PhotoShop, Adobe Illustrator, Netflix App, MATLAB, MCAFEE Antivirus.

5.3 Open Source Software

This is a type of software that has an open source code that is available to use for all users. It can be modified and shared to anyone for any purpose. Common examples of open source software used by programmers are: LibreOffice, PHP, GNU Image Manipulation Program (GIMP).

5.4 Closed Source Software

These are the types of software that are non-free for the programmers. For this software, the source code is the intellectual property of software publishers. It is also called "proprietary software" since only the original authors can copy, modify and share the software. The following are some of the most common examples of closed source software: .NET, Java, Microsoft Office, Adobe Photoshop.

5.5 Utility Software

Utility software is considered a subgroup of system software. They manage the performance of your hardware and application software installed on your computer, to ensure they work optimally. Some features of utility software include: antivirus and security software, file compressor, disk cleaner, disk defragmentation software, data backup software.

6. Conclusion

In conclusion, there can be multiple ways to classify different types of computer software. The software can be categorized based on the function they perform such as application software, system software, programming software, and driver software. They can also be classified based on different features such as the nature of source code, accessibility, and cost of usage.

New Words

software	[ˈsɒftweə]	*n.*软件
command	[kəˈmɑ:nd]	*n.*命令，指令
end-user	[end-ˈju:zə]	*n.*最终用户，终端用户
spreadsheet	[ˈspredʃi:t]	*n.*电子表格
analyze	[ˈænəlaɪz]	*vt.*分析
app	[æp]	*n.*计算机应用程序 *abbr.*应用(application)
non-essential	[nɒn-ɪˈsenʃl]	*adj.*不重要的，非本质的
requirement	[rɪˈkwaɪəmənt]	*n.*要求，需求；必要条件
quantitative	[ˈkwɒntɪtətɪv]	*adj.*定量的，数量（上）的
database	[ˈdeɪtəbeɪs]	*n.*数据库

suite	[swi:t]	*n.*（软件的）套件
mixture	[ˈmɪkstʃə]	*n.*混合，混杂
communication	[kəˌmju:nɪˈkeɪʃn]	*n.*通信；交流
browser	[ˈbraʊzə]	*n.*浏览器，浏览程序
email	[ˈi:meɪl]	*n.*电子邮件 *vt.*给……发电子邮件
background	[ˈbækgraʊnd]	*n.*后台，背景
desktop	[ˈdesktɒp]	*n.*桌面
tablet	[ˈtæblət]	*n.*平板计算机
develop	[dɪˈveləp]	*v.*开发
test	[test]	*v.*测试
debug	[ˌdi:ˈbʌg]	*vt.*调试，排除故障
bespoke	[bɪˈspəuk]	*adj.*定做的
facilitator	[fəˈsɪlɪteɪtə]	*n.*促进者，帮助者
compiler	[kəmˈpaɪlə]	*n.*编译器，编译程序
interpreter	[ɪnˈtɜ:prɪtə]	*n.*解释器，解释程序
assembler	[əˈsemblə]	*n.*汇编程序
editor	[ˈedɪtə]	*n.*编辑器，编辑软件，编辑程序
debugger	[ˌdi:ˈbʌgə]	*n.*调试器，调试程序
driver	[ˈdraɪvə]	*n.*驱动器，驱动程序
enable	[ɪˈneɪbl]	*vt.*使能够，使可能
assign	[əˈsaɪn]	*vt.*分配
standardize	[ˈstændədaɪz]	*vt.*使标准化；用标准校检
built-in	[bɪlt-ɪn]	*adj.*嵌入的；内置的
third-party	[ˈθɜ:dpɑ:tɪ]	*adj.*第三方的
installation	[ˌɪnstəˈleɪʃn]	*n.*安装
multiple	[ˈmʌltɪpl]	*adj.*多重的；多个的；多功能的
variant	[ˈveərɪənt]	*n.*变种，变异体；变形，变量 *adj.*不同的，相异的
separate	[ˈsepəreɪt]	*v.*（使）分开，分离；分割；划分；（使）分离，分散；隔开
subcategory	[ˈsʌbˈkætɪgərɪ]	*n.*亚类，子类
freeware	[ˈfri:weə]	*n.*免费软件
download	[ˌdaʊnˈləʊd]	*v.*下载
shareware	[ˈʃeəweə]	*n.*共享软件

trial	[ˈtraɪəl]	*adj.*试验的
antivirus	[ˈæntɪvaɪrəs]	*n.*抗病毒软件
available	[əˈveɪləbl]	*adj.*可利用的；可得到的；有效的
publisher	[ˈpʌblɪʃə]	*n.*发布者，发表者
proprietary	[prəˈpraɪətərɪ]	*adj.*专有的，专利的 *n.*所有权，所有物
optimally	['əptəməlɪ]	*adv.*最佳地
compressor	[kəmˈpresə]	*n.*压缩程序
accessibility	[əkˌsesə'bɪlɪtɪ]	*n.*可访问性

Phrases

engage with	接触；处理
instant messaging	即时通信
a collection of	一组，一些，一批
system software	系统软件
power up	加电，开机
low-level software	低级软件
machine language	机器语言
source code	源代码
machine code	机器代码
assembly code	汇编代码
be classified as	被归类为……，被认为……
plug into	接入（计算机系统），插入
open source software	开源软件
open source code	开源代码
closed source software	闭源软件
intellectual property	知识产权
original author	原作者
disk defragmentation software	磁盘碎片整理软件

Abbreviations

IDE (Integrated Development Environment)	集成开发环境
GNU (GNU's Not UNIX 的递归缩写)	革奴计划
GIMP (GNU Image Manipulation Program)	GNU 图像处理程序

Text A 参考译文

软件的不同类型

软件是一种计算机程序，提供执行用户命令的指令和数据。

1. 应用软件

作为技术用户，最常使用的是应用软件或“应用”。它们是高效的最终用户程序，可以帮助你执行任务。以下是一些应用程序软件示例，可让你执行特定工作：

- MS Excel：这是一个电子表格软件，可用于呈现和分析数据。
- Photoshop：它是 Adobe 的图片编辑应用程序软件。可以使用它来可视化地增强、分类和共享图片。
- Skype：这是一个在线通信应用程序，可用于视频聊天、语音呼叫和即时通信。

软件应用程序也称为非必需软件。它们可以根据用户要求在计算机上安装和操作。可以使用许多应用程序软件来执行不同的任务。随着技术的进步和用户需求的不断变化，此类应用的数量持续增加。可以将这些软件类型分为不同的组，如表 2-1 所示。

表 2-1　应用软件类型和示例

应用软件类型	示例
文字处理软件：用于创建文字表和键入文档等工具	Microsoft Word、WordPad、AppleWorks 和 Notepad
电子表格软件：用于计算定量数据的软件	Apple Numbers、Microsoft Excel 和 Quattro Pro
数据库软件：用于存储数据和排序信息	Oracle、MS Access 和 FileMaker Pro
应用程序套件：打包出售的相关程序的集合	OpenOffice、Microsoft Office
多媒体软件：用于混合音频、视频、图像和文本内容的工具	Real Player、Media Player
通信软件：用于连接系统并允许基于文本、音频和视频通信的工具	MS NetMeeting、IRC、ICQ
因特网浏览器：用于访问和查看网站	Netscape Navigator、MS Internet Explorer 和 Google Chrome
电子邮件程序：用于电子邮件发送的软件	Microsoft Outlook、Gmail、Apple Mail

2. 系统软件

系统软件可帮助用户、硬件和应用软件进行交互并协同工作。此类计算机软件是一个环境或平台，让其他软件和应用程序运行于其上。这就是系统软件对于管理整个计算机系统至关重要的原因。

首次打开计算机电源时，系统软件就被加载到内存中。与应用程序软件不同，最终用户不会使用系统软件。当使用其他应用程序软件时，它仅在最基本级别的设备后台运行。这就是系统软件也被称为“低级软件”的原因。

操作系统是系统软件的一个示例。所有类似计算机的设备都在操作系统上运行，包括台式机、笔记本计算机、智能手机和平板计算机等。这是一个操作系统示例的列表。让我们看一下，可能会发现一些熟悉的系统软件名称：

- 对于台式机、笔记本计算机和平板计算机：Microsoft Windows、Mac（用于 Apple 设备）和 Linux。
- 除操作系统外，有些人还将编程软件和驱动程序软件归类为系统软件。

3. 编程软件

编程软件是用于编写、开发、测试和调试其他软件（包括应用程序和系统软件）的程序。例如，对于在定制软件开发公司工作的人来说，这种类型的软件将使他们的生活更轻松、更高效。

编程软件被软件程序员用作转换器程序。它们是用于将编程语言（即 Java、C ++、Python、PHP、BASIC 等）转换为机器语言代码的辅助软件。转换器可以是编译器、解释器和汇编器。可以将编译器理解为将整个源代码转换为机器代码并执行的程序。当程序逐行运行时，解释器将运行源代码。汇编器将基本的计算机指令（汇编代码）转换为机器代码。

不同的编程语言编辑器、调试器、编译器和集成开发环境（IDE）是编程软件的示例。

4. 驱动程序软件

驱动程序软件通常被分入系统软件的类型。它们操作和控制插入计算机的设备和外围设备。驱动程序很重要，因为它们使设备能够执行其指定的任务。它们通过为硬件或设备转换操作系统的命令、分配任务来实现此目的。因此，与计算机连接的每个设备都需要至少一个设备驱动程序才能运行。

由于存在数千种设备，因此驱动程序通过允许其用标准化语言进行通信，使系统软件的工作更加轻松。你可能熟悉的一些驱动程序软件示例：打印机驱动程序、鼠标驱动程序。

通常，操作系统默认内置了鼠标、键盘和打印机的驱动程序。它们通常不需要第三方安装。但是对于某些高级设备，可能需要在外部安装驱动程序。此外，如果使用多个操作系统（例如 Linux、Windows 和 Mac），则每个操作系统都支持不同的驱动程序变体。对于它们，需要为每个设备编制单独的驱动程序。

5. 另一种软件分类

让我们讨论软件的另外 5 个子类别，并使用流行的软件示例来了解它们。

5.1　免费软件

免费软件指可以免费使用的任何软件。可以通过互联网免费下载和安装它们。免费软件的一些著名示例是：Google 浏览器、Skype、Instagram、Snapchat、Adobe Reader。

尽管它们都属于应用程序或最终用户软件的类别，但它们可以进一步归为免费软件，因为它们免费供用户使用。

5.2 共享软件

另一方面，共享软件是付费应用程序，但是在有限的“试用期”内免费提供。你可以在试用期内免费使用该软件，但是在试用期结束后，系统会要求购买该软件以继续使用。共享软件使你可以在实际投资购买软件之前对其进行测试。共享软件常见的一些示例包括：Adobe PhotoShop、Adobe Illustrator、Netflix App、MATLAB、McAfee Antivirus。

5.3 开源软件

这是一种所有用户都可使用的开源代码软件。无论出于何种目的，任何人都可对其进行修改并分享给其他人。程序员使用的开源软件的常见示例包括：LibreOffice、PHP、GNU 图像处理程序（GIMP）。

5.4 闭源软件

这类软件程序员不可免费获得。对于此软件，源代码是软件发行者的知识产权。它也被称为“专有软件”，因为只有原始作者才能复制、修改和分享该软件。以下是闭源软件的一些最常见示例：.NET、Java、Microsoft Office、Adobe Photoshop。

5.5 实用软件

实用程序软件被视为系统软件的子类。它们管理安装在计算机上的硬件和应用程序软件，以确保它们以最佳状态工作。实用程序软件的某些功能包括：防病毒和安全软件、文件压缩器、磁盘清理器、磁盘碎片整理软件及数据备份软件。

6. 结论

总之，可以有多种方法对不同类型的计算机软件进行分类。可以根据软件执行的功能将其分类，例如应用程序软件、系统软件、编程软件和驱动程序软件。还可以根据不同的特征对它们进行分类，例如源代码的性质、可访问性和使用成本。

Text B

The Seven Phases of the System Development Life Cycle

扫码听课文

The system development life cycle enables users to transform a newly-developed project into an operational one.

The system development life cycle, "SDLC" for short, is a multistep, iterative process, structured in a methodical way. This process is used to model or provide a framework for technical and non-technical activities to deliver a quality system which meets or exceeds a business's expectations or manage decision-making progression.

Traditionally, the systems development life cycle consisted of five stages. That has now increased to seven phases. Increasing the number of steps helps systems analysts to define clearer actions to achieve specific goals.

Similar to a project life cycle (PLC), the SDLC uses a systems approach to describe a process. It is often used and followed when there is an IT or IS project under development.

The SDLC highlights different stages of the development process. The life cycle approach is used so users can see and understand what activities are involved within a given step. It is also used to let them know that at any time, steps can be repeated or a previous step can be reworked when needing to modify or improve the system.

1. Planning

This is the first phase in the systems development process. It identifies whether or not there is the need for a new system to achieve a business's strategic objectives. This is a preliminary plan (or a feasibility study) for a company's business initiative to acquire the resources to build on an infrastructure to modify or improve a service. The company might be trying to meet or exceed expectations for their employees, customers and stakeholders too. The purpose of this step is to find out the scope of the problem and determine solutions. Resources, costs, time, benefits and other items should be considered at this stage.

2. Systems Analysis and Requirements

The second phase is where businesses will work on the source of their problem or the need for a change. In the event of a problem, possible solutions are submitted and analyzed to identify the best fit for the ultimate goals of the project. This is where teams consider the functional requirements of the project or solution. It is also where system analysis takes place — analyzing the needs of the end users to ensure the new system can meet their expectations. Systems analysis is vital in determining what a business's needs are, as well as how they can be met, who will be responsible for individual pieces of the project, and what sort of timeline should be expected.

There are several tools businesses can use that are specific to the second phase. They include:

- CASE (computer aided systems/software engineering)
- Requirements gathering
- Structured analysis

3. Systems Design

The third phase describes, in detail, the necessary specifications, features and operations that will satisfy the functional requirements of the proposed system which will be in place. This is the step for end users to discuss and determine their specific business information needs for the proposed system. It's during this phase that they will consider the essential components (hardware and/or software) structure (networking capabilities), processing and procedures for the system to accomplish its objectives.

4. Development

The fourth phase is when the real work begins, in particular, when a programmer, network engineer and/or database developer are brought on to do the major work on the project. This work includes using a flow chart to ensure that the process of the system is properly organized. The development phase marks the end of the initial section of the process. Additionally, this phase signifies the start of production. The development stage is also characterized by instillation and change. Focusing on training can be a huge benefit during this phase.

5. Integration and Testing

The fifth phase involves systems integration and system testing of programs — normally carried out by a quality assurance (QA) professional — to determine if the proposed design meets the initial set of business goals. Testing may be repeated, specifically to check for errors, bugs and interoperability. This testing will be performed until the end user finds it acceptable. Another part of this phase is verification and validation, both of which will help ensure the program's successful completion.

6. Implementation

The sixth phase is when the majority of the code for the program is written. Additionally, this phase involves the actual installation of the newly-developed system. This step puts the project into production by moving the data and components from the old system and placing them in the new system via a direct cutover. While this can be a risky and complicated move, the cutover typically happens during off-peak hours, thus minimizing the risk. Both system analysts and end-users should now see the realization of the project that has implemented changes.

7. Operations and Maintenance

The seventh and final phase involves maintenance and regular required updates. This step is when end users can fine-tune the system, if they wish, to boost performance, add new capabilities or meet additional user requirements.

If a business determines a change is needed during any phase of the SDLC, the company might have to proceed through all the above life cycle phases again. The life cycle approach of any project is a time-consuming process. Even though some steps are more difficult than others, none are to be overlooked. An oversight could prevent the entire system from functioning as planned.

New Words

development	[dɪˈveləpmənt]	*n.*开发，发展，进化
multistep	[ˈmʌltɪstep]	*adj.*多步的，多级的
iterative	[ˈɪtərətɪv]	*adj.*重复的，反复的，迭代的 *n.*反复体
methodical	[məˈθɒdɪkl]	*adj.*有方法的；有条不紊的
framework	[ˈfreɪmwɜːk]	*n.*构架；框架；（体系的）结构
activity	[ækˈtɪvɪtɪ]	*n.*活动
deliver	[dɪˈlɪvə]	*vt.*发表；交付 *vi.*投递；传送
exceed	[ɪkˈsiːd]	*vt.*超过；超越；胜过 *vi.*突出，领先
progression	[prəˈgreʃn]	*n.*（事件的）连续；一系列；发展，进展
achieve	[əˈtʃiːv]	*vt.*取得；获得；实现
repeated	[rɪˈpiːtɪd]	*adj.*反复的，再三的，重复的
rework	[ˌriːˈwɜːk]	*vt.*重做，再做；修订
plan	[plæn]	*n.*计划；打算
strategic	[strəˈtiːdʒɪk]	*adj.*战略性的，至关重要的
preliminary	[prɪˈlɪmɪnərɪ]	*adj.*初步的，初级的；预备的；开端的 *n.*准备工作；初步措施
feasibility	[ˌfiːzəˈbɪlɪtɪ]	*n.*可行性，可能性，现实性
initiative	[ɪˈnɪʃətɪv]	*n.*主动性；主动权 *adj.*自发的；创始的
customer	[ˈkʌstəmə]	*n.*顾客，客户
solution	[səˈluːʃn]	*n.*解决方案
requirement	[rɪˈkwɑɪəmənt]	*n.*要求，需求
submit	[səbˈmɪt]	*vt.*提交，呈送
identify	[ɑɪˈdentɪfaɪ]	*vt.*识别；确定
ultimate	[ˈʌltɪmət]	*adj.*最后的；极限的 *n.*终极

timeline	[ˈtaɪmlaɪn]	*n.*时间轴，时间表
satisfy	[ˈsætɪsfaɪ]	*vt.*符合，达到（要求、规定、标准等）
		*vi.*使满足或足够
essential	[ɪˈsenʃl]	*adj.*基本的；必要的；本质的
		*n.*必需品；基本要素
accomplish	[əˈkʌmplɪʃ]	*vt.*完成；达到（目的）
signify	[ˈsɪgnɪfaɪ]	*vt.*表示……的意思；意味；预示
		*vi.*具有重要性，要紧
integration	[ˌɪntɪˈgreɪʃn]	*n.*集成，整合
professional	[prəˈfeʃənl]	*adj.*专业的；职业的
		*n.*专业人士
determine	[dɪˈtɜ:mɪn]	*vt.*决定，确定；判定
bug	[bʌg]	*n.*缺陷，瑕疵
interoperability	['ɪntərɒpərə'bɪlɪtɪ]	*n.*互用性，协同工作的能力
verification	[ˌverɪfɪ'keɪʃn]	*n.*证明；证实
validation	[ˌvælɪ'deɪʃn]	*n.*确认；有效；校验
implementation	[ˌɪmplɪmen'teɪʃn]	*n.*执行，履行；实施，贯彻；生效；完成
involve	[ɪnˈvɒlv]	*vt.*包含；涉及，使参与
actual	[ˈæktʃuəl]	*adj.*真实的；实际的
cutover	['kʌtˌəʊvə]	*v.*切换，转换
off-peak	[ˈɔ:fˈpi:k]	*adj.*非高峰的
realization	[ˌri:əlaɪˈzeɪʃn]	*n.*实现；认识，领会
operation	[ˌɒpəˈreɪʃn]	*n.*运行，操作，经营
maintenance	[ˈmeɪntənəns]	*n.*维护；维修；保养
regular	[ˈregjʊlə]	*adj.*有规律的，规则的；定期的
fine-tune	[faɪn-tju:n]	*vt.*调整，对进行微调
update	[ˌʌpˈdeɪt]	*vt.*更新
boost	[bu:st]	*vt.*促进，提高
		*n.*提高，增加
time-consuming	[taɪmkən'sju:mɪŋ]	*adj.*费时的，旷日持久的
overlook	[ˌəʊvəˈlʊk]	*vt.*忽视，忽略
oversight	[ˈəʊvəsaɪt]	*n.*疏忽；失察

Phrases

approach to	接近
feasibility study	可行性研究
systems analysis	系统分析
requirements gathering	需求收集，需求采集
structured analysis	结构化分析
in detail	详细地
network engineer	网络工程师
database developer	数据库开发人员
flow chart	流程图
focus on	致力于；使聚焦于
systems integration	系统集成

Abbreviations

SDLC (System Development Life Cycle)	系统开发生命周期
PLC (Project Life Cycle)	项目生命周期
IS (Information System)	信息系统
CASE (Computer Aided Systems/Software Engineering)	计算机辅助系统/软件工程
QA (Quality Assurance)	质量保证

Text B 参考译文

系统开发生命周期的7个阶段

系统开发生命周期使用户可以将新开发的项目转换为可操作的项目。

系统开发生命周期（简称 SDLC）是一个多步骤的迭代过程，以系统的方式进行构造。此过程用于为技术和非技术活动建模或提供框架，以提供满足或超过企业期望或管理决策进度的质量体系。

传统的系统开发生命周期包括 5 个阶段。现在已增加到 7 个阶段。增加步骤数量有助于系统分析人员定义更清晰的操作从而实现特定目标。

与项目生命周期（PLC）相似，SDLC 使用系统方法来描述过程。当正在开发 IT 或 IS 项目时，通常会使用并遵循它。

SDLC 突出了开发过程的不同阶段。它使用了生命周期方法，因此用户可以查看和了解给定步骤中涉及哪些活动。还可以让他们知道，在需要修改或改进系统时，可以随时重复步骤或可以重新执行上一步。

1. 规划

这是系统开发过程的第一阶段。它确定是否需要新系统来实现业务的战略目标。这是公司业务计划的一项初步计划（或可行性研究），该计划旨在获取在基础设施上构建修改或改善服务的资源。该公司也可能试图达到或超过其员工、客户和利益相关者的期望。此步骤的目的是找出问题的范围并确定解决方案。在此阶段应考虑资源、成本、时间、收益和其他项目。

2. 系统分析与需求

第二阶段是企业将解决问题的根源或变更需求的阶段。出现问题时，将提交并分析可能的解决方案，以确定最适合项目最终目标的解决方案。此时，团队考虑项目或解决方案的功能要求。也进行系统分析——分析最终用户的需求，以确保新系统能够满足他们的期望。系统分析对于确定业务需求、如何满足需求、由谁来负责项目的各个部分以及预期的时间表至关重要。

企业可以使用几种特定于第二阶段的工具。它们包括：

- CASE（计算机辅助系统/软件工程）。
- 需求收集。
- 结构化分析。

3. 系统设计

第三阶段详细描述了必要的规范、特征和操作，这些规范、特征和操作需满足未来要实施的拟定系统的功能要求。这个步骤中，最终用户讨论和确定他们对拟定系统的特定业务信息需求。在此阶段，他们将考虑系统实现其目标所需的基本组件（硬件和/或软件）的结构（网络功能）、处理过程和步骤。

4. 开发

第四阶段是真正的工作开始的时候，尤其是当程序员、网络工程师和/或数据库开发人员参与项目的主要工作时。这项工作包括使用流程图来确保正确组织系统过程。开发阶段标志着该系统开发初始部分的结束。此外，此阶段表示生产开始。开发阶段的特征还在于灌输和变革。在此阶段，专注于培训可以带来巨大的好处。

5. 集成与测试

第五阶段涉及系统集成和程序的系统测试，通常由质量保证（QA）专业人员执行，以确定拟定的设计是否满足最初的业务目标。测试可以重复进行，尤其是检查错误、缺陷和互用性。测试可一直执行，直到最终用户接受为止。此阶段的另一部分是验证和确认，这两者将有助于确保程序的成功完成。

6. 实施

第六阶段是编写程序的大部分代码的时候。另外，此阶段涉及新开发系统的实际安装。此步骤将数据和组件从旧系统中移出并直接转换到新系统中，使项目投入生产。尽管这可能是一个冒险且很复杂的操作，但切换过程通常会在非高峰时段进行，从而将风险降到最低。系统分析人员和最终用户现在都应该看到已经实施更改的项目的实现。

7. 运营与维护

第七阶段也是最后阶段，涉及维护和定期更新。此步骤让最终用户可以根据需要微调系统以提高性能、添加新功能或满足其他用户要求。

如果企业确定在 SDLC 的任何阶段需要进行更改，则该企业可能不得不重复上述生命周期的各个阶段。任何项目的生命周期方法都是一个耗时的过程。即使某些步骤比其他步骤更困难，但也不应忽略任何步骤。疏忽可能会阻碍整个系统按计划运行。

Exercises

[Ex. 1] Answer the following questions according to Text A.

1. What are some examples of application software that allow you to do specific work?
2. What is Word processing software? What are the examples?
3. What does system software do? Why is it essential in managing the whole computer system?
4. What is programming software?
5. Why are drivers important?
6. What are some examples of driver software that you may be familiar with?
7. What is freeware software? What are some well-known examples of freeware?
8. What does shareware allow you to do? What are some examples of shareware that you must be familiar with?
9. What are common examples of open source software used by programmers?
10. What are some of the most common examples of closed source software?

[Ex.2] Fill in the following blanks according to Text B.

1. The system development life cycle, "SDLC" for short, is a ____________, iterative

process, structured in a ___________ way. This process is used to model or provide a framework for technical and non-technical activities to deliver a ______________ which meets or exceeds a business's expectations or manage ______________.

2. Traditionally, the systems development life cycle consisted of ___________ stages. That has now increased to ___________ phases.
3. The life cycle approach is used so users can see and understand what activities are involved ____________. It is also used to let them know that ____________, steps can be repeated or a previous step ____________ when needing to ____________________.
4. Planning is the first phase ____________________. It identifies whether or not there is the need for a new system to ____________________.
5. There are several tools businesses can use that are specific to the second phase. They include:

____________________, ____________________ and ____________________.

6. The third phase describes, in detail, ____________________, ____________ and ____________ that will satisfy the functional requirements of the proposed system which will be in place.
7. The development phase marks ____________________ of the process. Additionally, this phase signifies ____________________. The development stage is also characterized by____________________.
8. The fifth phase involves ____________________ and ____________________ — normally carried out by a quality assurance (QA) professional — to determine if the proposed design meets ____________________.
9. The sixth phase is when the majority of the code for____________ is written. Additionally, this phase involves ____________________ of the newly-developed system.
10. The life cycle approach of any project is ____________________. Even though some steps are ______________ than others, none are to be overlooked. An oversight could prevent the entire system from ______________.

[Ex. 3] Translate the following terms or phrases from English into Chinese and vice versa.

1. assembly code	1. ____________________
2. machine language	2. ____________________
3. source code	3. ____________________
4. flow chart	4. ____________________
5. systems analysis	5. ____________________
6. *n.*抗病毒软件	6. ____________________

7. n.汇编程序 7. ____________
8. n.调试器，调试程序 8. ____________
9. n.解释器，解释程序 9. ____________
10. n.缺陷，瑕疵 10. ____________

[Ex. 4] Translate the following sentences into Chinese.

Responsibilities and Tasks of Software Developers

1. Primary Responsibilities

Here's a non-exhaustive list of common tasks software developers are expected to complete.

- Creating and developing new software. Researching users' requirements, designing and writing new software, and testing newly designed software.
- Evaluating new and existing software systems. Designing testing plans for newly developed software, performing QA testing on software systems, finding faults in software systems, and correcting faults found in software systems.
- Improving existing software systems. Analyzing users' requirements and suggestions, creating solutions for existing issues, and implementing these solutions.
- Performing maintenance to existing systems by monitoring and correcting defects. Writing code (e.g. HTML, PHP, XML) for new software and updates. Running code to test efficiency, rewriting code to correct errors, and running tests again until code is error free.
- Writing operational manuals and systems specifications.
- Working in tandem with other staff members such as project managers, graphic designers, other developers, database administrators, and sales and marketing employees. Consulting with clients or project managers on the progress of developing software to check for possible improvements, suggestions, or requirements.
- Writing reports on project progress.

2. Daily Tasks

- Meeting with clients and project managers to design and develop new software.
- Establishing parameters and designing the architecture of new software.
- Designing, writing, reading, testing, and correcting code for new software.
- Running QA testing and searching for bugs in developing software.
- Reporting to clients and project managers on the development of new software.
- Testing and implementing software updates and improvements when necessary.
- Writing documentation for new and updated software.

[Ex. 5] Fill in the blanks with the words given below.

new	interact	data	complex	software
programs	designing	project	testing	create

Software Developers

Software developers often work for computer firms and manufacturers. Their main role is to ____1____ the foundations for operative systems on which computer programmers work. They design, write, and test code for new systems and ____2____ to ensure efficiency. Software developers also run diagnostic ____3____ and quality assurance (QA) testing on existing projects before launching them to certify effectiveness.

A software developer is involved in all the process related to creating and ____4____ new systems; from initial planning, to establishing parameters, designing, writing, coding, encrypting, and ____5____. This process is usually undertaken by a team of software developers, with each member carrying out a particular step of the process and a supervisor overseeing the entire ____6____.

The work of a software developer may sometimes overlap with that of a database administrator. Many systems have to ____7____ in one way or another with data management systems, so it is the responsibility of the software developer to ensure that both systems are compatible. Some software developers can do this by themselves if they possess enough knowledge on ____8____ management systems and software.

Software developers often use several programming languages, their job is often very ____9____ and it involves advanced knowledge in computer science and mathematics. Their field is constantly evolving and ____10____ technologies and advancements are made every day, so they must be in a constant state of learning and self-improvement.

Online Resources

二维码	内 容
	计算机专业常用语法（2）：状语从句
	在线阅读（1）：The Different Types of Software Testing（软件测试的不同类型）
	在线阅读（2）：Top 4 Software Development Methodologies（4 种常用的软件开发方法）

Unit 3

Operating System

Text A

Operating System

An operating system acts as an intermediary between the user of a computer and computer hardware. The purpose of an operating system is to provide an environment in which a user can execute programs in a convenient and efficient manner.

An operating system is a software that manages the computer hardware. The hardware must provide appropriate mechanisms to ensure the correct operation of the computer system and to prevent user programs from interfering with the proper operation of the system.

1. Definition

- An operating system is a program that controls the execution of application programs and acts as an interface between the user of a computer and the computer hardware.
- A more common definition is that the operating system is the program running at all times on the computer (usually called the kernel), with all else being application programs.
- An operating system is concerned with the allocation of resources and services, such as memory, processors, devices, and information. The operating system correspondingly includes programs to manage these resources, such as a traffic controller, a scheduler, memory management module, I/O programs, and a file system.

2. Functions of Operating System

Operating system performs three functions:

- Convenience: An OS makes a computer more convenient to use.
- Efficiency: An OS allows the computer system resources to be used in an efficient manner.
- Ability to evolve: An OS should be constructed in such a way as to permit the effective development, testing and introduction of new system functions at the same time without interfering with service.

3. Types of Operating Systems

An operating system performs all the basic tasks like managing file, process and memory. It acts as manager of all the resources. It becomes an interface between user and machine.

3.1 Batch Operating System

This type of operating system does not interact with the computer directly.

Advantages of batch operating system:

- Multiple users can share the batch systems.
- The idle time for batch system is very less.
- It is easy to manage large work repeatedly in batch systems.

Disadvantages of batch operating system:

- Batch systems are hard to debug.
- It is sometime costly.
- The other jobs will have to wait for an unknown time if any job fails.

3.2 Time-Sharing Operating System

Each task is given some time to execute, so that all the tasks work smoothly. Each user gets time of CPU as they use single system. These systems are also known as multitasking systems. The task can be from single user or from different users as well. The time that each task gets to execute is called time slice. After this time interval is over, OS switches over to next task.

Advantages of time-sharing OS:

- Each task gets an equal opportunity.
- Less chances of duplication of software.
- CPU idle time can be reduced.

Disadvantages of time-sharing OS:

- Reliability problem.
- One must take care of security and integrity of user programs and data.
- Data communication problem.

3.3 Distributed Operating System

This type of operating system is a recent advancement in the world of computer technology and is being widely accepted all over the world and with a great pace. Various autonomous interconnected computers communicate with each other using a shared communication network. Independent systems possess their own memory unit and CPU. These are referred to as loosely coupled systems or distributed systems. The processors of these systems differ in size and function. The major benefit of working with this type of operating system is that it is always possible that one user can access the files or software which is not actually present on his system but on some other system connected within this network.

Advantages of distributed operating system:

- Failure of one will not affect the other network communication, as all systems are independent from each other.
- Electronic mail increases the data exchange speed.
- Since resources are being shared, computation is highly fast and durable.
- Load on host computer is reduced.
- These systems are easily scalable as many systems can be easily added to the network.
- Delay in data processing is reduced.

Disadvantages of distributed operating system:

- Security problem due to sharing.
- Some messages can be lost in the network system.
- Overloading is another problem in distributed operating systems.
- If there is a database connected on local system and many users accessing that database through remote or distributed way then performance becomes slow.
- The databases in network operating is more difficult to administrate than single user system.

3.4 Network Operating System

These systems run on a server and provide the capability to manage data, users, groups, security, applications, and networking functions. This type of operating system allows shared access of files, printers, security, applications, and other networking functions over a small private network. One more important aspect of network operating systems is that all the users are well aware of the underlying configuration and of all other users within the network, their individual connections etc., and that's why these computers are popularly known as tightly coupled systems.

Advantages of network operating system:

- Highly stable centralized servers.
- Security concerns are handled through servers.
- New technologies and hardware upgradation are easily integrated to the system.
- Server access are possible remotely from different locations and types of systems.

Disadvantages of network operating system:

- Servers are costly.
- User has to depend on central location for most operations.
- Maintenance and updates are required regularly.

3.5 Real-Time Operating System

This type of OS serves the real-time systems. The time interval required to process and respond to inputs is very small. This time interval is called response time.

Real-time systems are used when time requirements are very strict like missile systems, air traffic control systems, robots etc.

There are two types of real-time operating systems.

- Hard real-time systems: These OSs are meant for the applications that time constraints are very strict and even the shortest possible delay is not acceptable. These systems are built for saving life like automatic parachutes or air bags which are required to be readily available in case of any accident. Virtual memory is almost never found in these systems.
- Soft real-time systems: These OSs are for applications that for time-constraint is less strict.

Advantages of RTOS:

- Maximum consumption: Maximum utilization of devices and system, thus more output from all the resources.
- Task shifting: Time assigned for shifting tasks in these systems is very less. For example, in older systems it takes about 10 micro seconds in shifting one task to another and in latest systems it takes 3 micro seconds.
- Focus on application: Focus on running applications and less importance to applications which are in queue.
- Real time operating system in embedded system: Since the size of programs is small, RTOS can also be used in embedded systems like in transport and others.
- Error free: These types of systems are error free.
- Memory allocation: Memory allocation is best managed in this type of system.

Disadvantages of RTOS:

- Use heavy system resources: Sometimes the system resources are not so efficient and they are expensive as well.
- Complex algorithms: The algorithms are very complex and difficult for the designer to write on.
- Device driver and interrupt signals: It needs specific device drivers and interrupt signals to response earliest to interrupts.
- Thread priority: It is not good to set thread priority as these systems are very less prone to switching tasks.

Examples of real-time operating systems are scientific experiments, medical imaging systems, industrial control systems, weapon systems, robots, air traffic control systems, etc.

New Words

intermediary	[ˌɪntəˈmi:dɪərɪ]	*adj.*中间人的；居间的；媒介的 *n.*中间人；媒介
environment	[ɪnˈvaɪrənmənt]	*n.*环境，外界
convenient	[kənˈvi:nɪənt]	*adj.*方便的
prevent	[prɪˈvent]	*v.*防止，阻止

definition	[ˌdefɪˈnɪʃn]	*n.*定义；规定；解释
kernel	[ˈkɜ:nl]	*n.*核；核心；要点
allocation	[ˌæləˈkeɪʃn]	*n.*配给，分配
correspondingly	[ˌkɒrəˈspɒndɪŋlɪ]	*adv.*相对地，比照地
controller	[kənˈtrəʊlə]	*n.*管理者；控制者；（机器的）控制器
scheduler	['ʃedju:lə]	*n.*调度程序，日程安排程序
module	[ˈmɒdju:l]	*n.*模块；组件
convenience	[kənˈvi:nɪəns]	*n.*方便，便利
efficient	[ɪˈfɪʃnt]	*adj.*有效率的；（直接）生效的；能干的
ability	[əˈbɪlɪtɪ]	*n.*能力，才能
evolve	[ɪˈvɒlv]	*vt.*使发展；使进化 *vi.*发展
introduction	[ˌɪntrəˈdʌkʃn]	*n.*采用，引进
perform	[pəˈfɔ:m]	*v.*执行，履行，运行
smoothly	[ˈsmu:ðlɪ]	*adv.*平滑地；流畅地；平稳地
opportunity	[ˌɒpəˈtju:nɪtɪ]	*n.*机会
duplication	[ˌdju:plɪ'keɪʃn]	*n.*复制；重复；成倍
reliability	[rɪˌlaɪə'bɪlɪtɪ]	*n.*可靠，可靠性
integrity	[ɪnˈtegrɪtɪ]	*n.*完整性
advancement	[ədˈvɑ:nsmənt]	*n.*前进，进步；提升，升级
autonomous	[ɔ:ˈtɒnəməs]	*adj.*自治的，有自主权的
interconnect	[ˌɪntəkəˈnekt]	*v.*互相连接，互相联系
possess	[pəˈzes]	*vt.*拥有；掌握
distributed	[dɪs'trɪbju:tɪd]	*adj.*分布式的
independent	[ˌɪndɪˈpendənt]	*adj.*自主的，不关联的
exchange	[ɪksˈtʃeɪndʒ]	*n.*交换；交易 *vt.*交换，互换，调换 *vi.*交换，替换
computation	[ˌkɒmpjʊˈteɪʃn]	*n.*计算
durable	[ˈdjʊərəbl]	*adj.*耐用的，耐久的；持久的；长期的 *n.*耐用品，耐久品
host	[həʊst]	*n.*主机
scalable	['skeɪləbl]	*adj.*可升级的
message	[ˈmesɪdʒ]	*n.*信息；消息 *v.*给……发消息；给……留言

overload	[ˌəʊvəˈləʊd] [ˈəʊvələʊd]	*vt.*使负担太重；使超载；超过负荷 *n.*过量，超负荷
administrate	[əd'mɪnɪstreɪt]	*v.*管理，支配
aspect	[ˈæspekt]	*n.*方面；面貌
configuration	[kənˌfigəˈreɪʃn]	*n.*布局，构造；配置
tightly	[ˈtaɪtlɪ]	*adv.*紧紧地，坚固地，牢固地
centralized	[sentrəlaɪzd]	*adj.*集中的
upgradation	[ˌʌpgreɪˈdeɪʃn]	*n.*升级，改善，提高
remotely	[rɪˈməʊtlɪ]	*adv.*远程地
real-time	['ri:l'taɪm]	*adj.*（计算机）即时处理的，实时的
interval	[ˈɪntəvl]	*n.*间隔
respond	[rɪˈspɒnd]	*vi.*做出反应，响应，回应
strict	[strɪkt]	*adj.*精确的；严格的
robot	[ˈrəʊbɒt]	*n.*机器人；遥控装置；自动机
constraint	[kənˈstreɪnt]	*n.*强制；限制；约束
acceptable	[əkˈseptəbl]	*adj.*可接受的；令人满意的
accident	[ˈæksɪdənt]	*n.*意外事件；事故
consumption	[kənˈsʌmpʃn]	*n.*消耗
utilization	[ˌju:təlaɪ'zeɪʃn]	*n.*利用，使用，效用
thread	[θred]	*n.*线程
prone	[prəʊn]	*adj.*易于……的；有……倾向的

Phrases

operating system	操作系统
act as	担当……，起……的作用；充当
interfere with	干扰
be concerned with ...	涉及……，与……有关
file system	文件系统
be constructed in ...	用……构造
batch operating system	批处理操作系统
multiple user	多用户
idle time	空闲时间
time-sharing operating system	分时操作系统
single system	单系统
multitasking system	多任务系统

single user	单用户
time slice	时间片
switch over	切换，变换，转换
take care of	照顾；对付；抵消
distributed operating system	分布式操作系统
electronic mail	电子邮件
single user system	单用户系统
network operating system	网络操作系统
private network	私有网络
tightly coupled system	紧耦合系统
response time	响应时间
air traffic control system	航空交通管制系统；空中交通管制系统
virtual memory	虚拟内存
micro second	微秒
embedded system	嵌入式系统
interrupt signal	中断信号
thread priority	线程优先权

Abbreviations

I/O (Input/Output)	输入/输出
RTOS (Real-Time Operating System)	实时操作系统

Text A 参考译文

操 作 系 统

操作系统充当计算机用户和计算机硬件之间的中介。操作系统的目的是提供一种环境，在该环境中用户可以方便且有效地执行程序。

操作系统是管理计算机硬件的软件。硬件必须提供适当的机制，以确保计算机系统的正确运行并防止用户程序干扰系统的正常运行。

1. 定义

- 操作系统是控制应用程序执行的程序，并且充当计算机用户和计算机硬件之间的接口。

- 更为常见的定义是，操作系统是始终在计算机上运行的程序（通常称为内核），而所有其他程序都是应用程序。
- 操作系统与资源和服务（例如内存、处理器、设备和信息）的分配有关。操作系统相应地包括用于管理这些资源的程序，例如流量控制器、调度程序、内存管理模块、I/O 程序和文件系统。

2. 操作系统的功能

操作系统执行三个功能：

- 便利性：操作系统使计算机更易于使用。
- 效率：操作系统让计算机系统资源被高效地使用。
- 进化能力：操作系统的构建方式应在不干扰服务的同时，允许有效地开发、测试和引入新的系统功能。

3. 操作系统的类型

操作系统执行所有基本任务，例如管理文件、进程和内存。它充当所有资源的管理者。它成为用户和机器之间的接口。

3.1 批处理操作系统

此类操作系统不会直接与计算机交互。

批处理操作系统的优点：

- 多个用户可以共享批处理系统。
- 批处理系统的空闲时间非常短。
- 在批处理系统中易于重复管理大型的工作。

批处理操作系统的缺点：

- 批处理系统难以调试。
- 有时很昂贵。
- 如果任何作业失败，其他作业将必须等待未知的时间。

3.2 分时操作系统

每个任务都有一定的执行时间，这样所有任务都能顺利进行。每个用户使用单个系统时都会获得 CPU 时间。这些系统也称为多任务系统。该任务可以来自单个用户，也可以来自不同用户。每个任务执行的时间称为时间片。在此时间间隔结束后，操作系统将切换到下一个任务。

分时操作系统的优势：

- 每个任务都有平等的机会。

- 复制软件的机会更少。
- 可以减少 CPU 空闲时间。

分时操作系统的缺点：

- 可靠性问题。
- 必须注意用户程序和数据的安全性和完整性。
- 数据通信问题。

3.3 分布式操作系统

这种类型的操作系统是计算机技术领域的最新进展，并且正在全世界范围内快速地被广泛接受。各种自主互联的计算机使用共享的通信网络相互通信。独立的系统拥有自己的存储单元和 CPU。这些被称为松耦合系统或分布式系统。这些系统的处理器在大小和功能上有所不同。使用这种类型的操作系统的主要好处是，用户总是可以访问自己的系统中实际上没有、但与该网络连接的其他系统上具有的文件或软件。

分布式操作系统的优点：

- 一个系统的故障不会影响另一网络的通信，因为所有系统都是相互独立的。
- 电子邮件可以提高数据交换速度。
- 由于共享资源，计算非常快速又持久。
- 减轻了主机上的负载。
- 这些系统易于扩展，因为许多系统可以轻松添加到网络中。
- 减少了数据处理的延迟。

分布式操作系统的缺点：

- 共享引起的安全问题。
- 某些消息可能会在网络系统中丢失。
- 重载是分布式操作系统中的另一个问题。
- 如果本地系统上已连接数据库，并且许多用户通过远程或分布式方式访问该数据库，则执行会变慢。
- 网络操作中的数据库比单用户系统难以管理。

3.4 网络操作系统

这些系统在服务器上运行，并提供管理数据、用户、组、安全性、应用程序和联网的功能。这种类型的操作系统允许通过小型专用网络共享访问文件、打印机、安全性、应用程序和其他网络功能。网络操作系统的另一个重要方面是，所有用户都清楚底层配置以及网络中的所有其他用户、它们的单独连接等，这就是这些计算机被普遍称为紧耦合系统的原因。

网络操作系统的优点：

- 高度稳定的集中式服务器。

- 通过服务器处理安全问题。
- 新技术和硬件升级很容易集成到系统中。
- 可以从不同位置和类型的系统远程访问服务器。

网络操作系统的缺点：

- 服务器价格昂贵。
- 对于大多数操作，用户必须依靠中央节点。
- 定期需要维护和更新。

3.5 实时操作系统

这种类型的操作系统服务于实时系统。处理和响应输入所需的时间间隔很小。该时间间隔称为响应时间。

当时间要求非常严格时，可以使用实时系统，例如导弹系统、空中交通管制系统、机器人等。

有两种实时操作系统：

- 硬实时系统：这些操作系统适用于时间限制非常严格，甚至最短的延迟都不可以接受的应用。这些系统是为挽救生命而设计的，例如自动降落伞或安全气囊，它们在发生任何事故时都必须随时可用。在这些系统中几乎找不到虚拟内存。
- 软实时系统：这些操作系统适用于时间限制不太严格的应用程序。

实时操作系统的优势：

- 消耗效率最大化：设备和系统的利用率最高，能够从所有资源获得更多输出。
- 任务转移：在这些系统中分配给转移任务的时间非常少。例如，在较旧的系统中，将一项任务转移到另一项任务大约需要 10 微秒，而在最新的系统中，则需要 3 微秒。
- 关注应用程序：关注正在运行的应用程序，而对排队的应用程序不太重视。
- 嵌入式系统中的实时操作系统：由于程序很小，因此实时操作系统也可以在嵌入式系统中使用，例如在传输系统和其他系统中。
- 无错误：这些类型的系统没有错误。
- 内存分配：在此类系统中内存分配最好管理。

实时操作系统的缺点：

- 使用大量的系统资源：有时系统资源效率不高，而且资源也很昂贵。
- 复杂的算法：算法非常复杂，设计人员难以编写。
- 设备驱动程序和中断信号：需要特定的设备驱动程序和中断信号来最早响应中断。
- 线程优先级：不便设置线程优先级，因为这些系统不太容易切换任务。

实时操作系统的示例是科学实验、医学成像系统、工业控制系统、武器系统、机器人、空中交通管制系统等。

Text B

What Is Linux

Linux is the best-known and most-used open source operating system. As an operating system, Linux is the software that sits underneath all of the other software on a computer, receiving requests from those programs and relaying these requests to the computer's hardware.

Here we use the term "Linux" to refer to the Linux kernel, and the set of programs, tools, and services that are typically bundled together with the Linux kernel to provide all of the necessary components of a fully functional operating system. Some people, particularly members of the Free Software Foundation, refer to this collection as GNU/Linux, because many of the tools included are GNU components. However, not all Linux installations use GNU components as a part of their operating system. Android, for example, uses a Linux kernel but relies very little on GNU tools.

1. How Does Linux Differ from Other Operating Systems

In many ways, Linux is similar to other operating systems you may have used before, such as Windows, OS X, or iOS. Like other operating systems, Linux has a graphical interface and types of software you are accustomed to using on other operating systems, such as word processing applications. In many cases, the software's creator may have made a Linux version of the same program you use on other systems. If you can use a computer or other electronic device, you can use Linux.

But Linux is different from other operating systems in many important ways. First, and perhaps most importantly, Linux is open source software. The code used to create Linux is free and available to the public to view, edit, and — for users with the appropriate skills — to contribute to.

Linux is also different in that, although the core pieces of the Linux operating system are generally common, there are many distributions of Linux, which include different software options. This means that Linux is incredibly customizable, because not just applications, such as word processors and web browsers, can be swapped out, Linux users can also choose core components, such as which system displays graphics and other user-interface components.

2. What Is the Difference Between UNIX and Linux

You may have heard of UNIX, which is an operating system developed in the 1970s at Bell Labs by Ken Thompson, Dennis Ritchie, and others. UNIX and Linux are similar in many ways. They both have similar tools for interfacing with the systems, programming tools, filesystem layouts, and other key components. However, UNIX is not free. Over the years, a number of different operating systems have been created that attempted to be "UNIX-like" or "UNIX-compatible," but Linux has been the most successful, far surpassing its predecessors in popularity.

3. Who Uses Linux

You're probably already using Linux, whether you know it or not. Depending on a user survey, between one- and two-thirds of the web pages on the Internet are generated by servers running Linux.

Companies and individuals choose Linux for their servers because it is secure, and you can receive excellent support from a large community of users, in addition to companies like Canonical, SUSE, and Red Hat, which offer commercial support.

Many of the devices you own, such as Android phones, digital storage devices, personal video recorders, cameras, wearables, and more, probably also run Linux. Even your car has Linux running under the hood.

4. Who "Owns" Linux

By virtue of its open source licensing, Linux is freely available to anyone. However, the trademark on the name "Linux" rests with its creator, Linus Torvalds. The source code for Linux is under copyright by its many individual authors, and licensed under the GPLv2 license. Because Linux has such a large number of contributors from across multiple decades of development, contacting each individual author and getting them to agree to a new license is virtually impossible, that Linux remaining licensed under the GPLv2 in perpetuity is all but assured.

5. How Was Linux Created

Linux was created in 1991 by Linus Torvalds, a then-student at the University of Helsinki. Torvalds built Linux as a free and open source alternative to Minix, another UNIX clone that was predominantly used in academic settings. He originally intended to name it "Freax", but the administrator of the server Torvalds used to distribute the original code named his directory "Linux" after a combination of Torvalds' first name and the word UNIX.

6. How Can I Contribute to Linux

Most of the Linux kernel is written in the C programming language, with a little bit of assembly and other languages sprinkled in. If you're interested in writing code for the Linux kernel itself, a good place to get started is in the Kernel Newbies FAQ, which will explain some of the concepts and processes you'll want to be familiar with.

But the Linux community is much more than the kernel, and it needs contributions from lots of other people besides programmers. Every distribution contains hundreds or thousands of programs that can be distributed along with it, and each of these programs, as well as the distribution itself, need a variety of people and skill sets to make them successful, including:

- Testers to make sure everything works well on different configurations of hardware and software, and to report the bugs when it does not.
- Designers to create user interfaces and graphics distributed with various programs.
- Writers who can create documentation, how-tos, and other important text distributed with software.
- Translators to take programs and documentation from their native languages and make them accessible to people around the world.
- Packagers to take software programs and put all the parts together to make sure they run flawlessly in different distributions.
- Disseminators to spread the word about Linux and open source in general.
- And of course developers to write the software itself.

7. How Can I Get Started Using Linux

There's some chance you're using Linux already and don't know it, but if you'd like to install Linux on your home computer to try it out, the easiest way is to pick a popular distribution that is designed for your platform (for example, laptop or tablet device) and give it a shot. Although there are numerous distributions available, most of the older, well-known distributions are good choices for beginners because they have large user communities that can help answer questions if you get stuck or can't figure things out. Popular distributions include Debian, Fedora, Mint, and Ubuntu, but there are still many others.

New Words

relay	[ˈriːleɪ]	*n.*传递；继电器 *vt.*转播，传达
service	[ˈsɜːvɪs]	*n.*服务
bundle	[ˈbʌndl]	*n.*捆 *vt.*额外免费提供（设备等），（尤指出售计算机时）赠送软件
foundation	[faʊnˈdeɪʃn]	*n.*基础；基金（会）
creator	[krɪˈeɪtə]	*n.*创造者，创作者
contribute	[kənˈtrɪbjuːt]	*v.*贡献出；出力
incredibly	[ɪnˈkredəblɪ]	*adv.*难以置信地，很，极为
customizable	[ˈkʌstəmaɪzəbl]	*adj.*可定制的，用户化的
compatible	[kəmˈpætəbl]	*adj.*兼容的，相容的
predecessor	[ˈpriːdɪsesə]	*n.*前任，前辈；原有事物，前身
popularity	[ˌpɒpjʊˈlærɪtɪ]	*n.*普及，流行

survey	[ˈsɜ:veɪ]	*vt.*调查
		*n.*调查（表）；测量
Android	[ˈændrɔɪd]	*n.*安卓操作系统
hood	[hʊd]	*n.*车篷；引擎罩
		*vt.*罩上；覆盖
licensing	['laɪsnsɪŋ]	*v.*批准，许可，颁发执照
trademark	[ˈtreɪdmɑ:k]	*n.*（注册）商标
copyright	[ˈkɒpɪraɪt]	*n.*版权，著作权
virtually	[ˈvɜ:tʃuəlɪ]	*adv.*实际上，实质上，事实上，几乎
perpetuity	[ˌpɜ:pəˈtju:ɪtɪ]	*n.*永久，永恒，永远
assured	[əˈʃʊəd]	*adj.*确定的
clone	[kləʊn]	*n.&v.*克隆
predominantly	[prɪˈdɒmɪnəntlɪ]	*adv.*占主导地位地；显著地；占优势地
concept	[ˈkɒnsept]	*n.*观念，概念，观点，思想
community	[kəˈmju:nətɪ]	*n.*社区；社会团体
skill	[skɪl]	*n.*技能，技巧
tester	[ˈtestə]	*n.*测试员
designer	[dɪˈzɑɪnə]	*n.*设计师；设计者
accessible	[əkˈsesəbl]	*adj.*可理解的；易接近的
flawlessly	['flɔ:lɪslɪ]	*adv.*无瑕地，完美地
disseminator	[dɪ'semɪneɪtə]	*n.*传播者
numerous	[ˈnju:mərəs]	*adj.*很多的，许多的

Phrases

open source	开源
receive from	从……获得，收到（某物）
Free Software Foundation	自由软件基金会
graphical interface	图形界面
be accustomed to ...	习惯于
electronic device	电子设备
be different from ...	不同于
web browser	网页浏览器，网络浏览器
Bell Lab	贝尔实验室
by virtue of ...	凭借……的力量，由于……
rest with	在于，取决于

a little bit	一点点
be familiar with	熟悉，认识
make sure	确保
native language	母语
get stuck	陷入僵局，抛锚，卡住
figure out	弄明白；解决；想出；计算出

Abbreviations

GPL (GNU General Public License)	GNU 通用公共授权
FAQ (Frequently Asked Questions)	常见问题，频繁问到的问题

Text B 参考译文

什么是 Linux

Linux 是最著名、使用最广泛的开源操作系统。作为操作系统，Linux 是一种软件，它位于计算机上所有其他软件的底层，可以接收来自这些程序的请求并将这些请求转发到计算机的硬件。

在这里，我们使用术语“Linux”来指代 Linux 内核以及通常与 Linux 内核捆绑在一起的程序、工具和服务集合，它们为全功能的操作系统提供了所有必要组件。有些人，特别是自由软件基金会的成员，将此集合称为 GNU/Linux，因为其中包含的许多工具都是 GNU 组件。但是，并非所有 Linux 安装都将 GNU 组件用作其操作系统的一部分。例如，Android 使用 Linux 内核，但几乎不依赖 GNU 工具。

1. Linux 与其他操作系统有何不同

在许多方面，Linux 类似于你以前使用过的其他操作系统（例如 Windows，OS X 或 iOS）。与其他操作系统一样，Linux 具有图形界面和你习惯在其他操作系统上使用的软件类型（例如文字处理应用程序）。在许多情况下，软件的创建者可能已经为你在其他系统上使用的同一程序制作了 Linux 版本。如果你会使用计算机或其他电子设备，则可以使用 Linux。

但是 Linux 在许多重要方面与其他操作系统有所不同。首先，也许是最重要的是，Linux 是开源软件。用于创建 Linux 的代码是免费的，可供公众查看、编辑，并且具有适当技能的用户也可为此做出贡献。

尽管 Linux 操作系统的核心部分很常见，但 Linux 仍然具有特色。Linux 有许多发行版本，其中包括不同的软件选项。这意味着 Linux 有极佳的可定制性，因为不仅可以替换应用程序（例如文字处理器和网页浏览器），Linux 用户还可以选择核心组件，例如显示图形的系统和其他用户界面组件。

2. UNIX 和 Linux 有什么区别

你可能已经听说过 UNIX，它是 Ken Thompson、Dennis Ritchie 等人在 20 世纪 70 年代于贝尔实验室开发的操作系统。UNIX 和 Linux 在许多方面都相似。它们都具有类似的工具，用于与系统、编程工具、文件系统布局和其他关键组件进行交互。但是，UNIX 不是免费的。多年来，已经创建了许多尝试成为“类 UNIX”或“UNIX 兼容”的不同操作系统，但是 Linux 一直是最成功的，远远超过了其前任。

3. 谁使用 Linux

无论你是否知道，你可能已经在使用 Linux。根据用户调查，因特网上三分之一至三分之二的网页是由运行 Linux 的服务器生成的。

公司和个人为其服务器选择 Linux 的原因在于，它是安全的，除了 Canonical、SUSE 和 Red Hat 等提供商业支持的公司之外，还可以从广大用户社区获得出色的支持。

你拥有的许多设备（例如 Android 手机、数字存储设备、个人录像机、照相机、可穿戴设备等）都有可能在运行 Linux。甚至你的汽车都在运行 Linux。

4. 谁“拥有”Linux

由于其开源许可，任何人都可以免费使用 Linux。但是，名称“Linux”的商标属于其创建者 Linus Torvalds。Linux 的源代码的版权属于许多个人作者，并已获得 GPLv2 许可。由于 Linux 在数十年的发展中出现了大量的贡献者，因此几乎不可能联系每个单独的作者并让他们同意新的许可，因此几乎可以肯定 Linux 仍在 GPLv2 下永久获得许可。

5. Linux 是如何创建的

Linux 是由 Linus Torvalds 于 1991 年创建，当时他是赫尔辛基大学的学生。Torvalds 将 Linux 构建为 Minix 的免费、开源的替代品，Minix 是 UNIX 的一个克隆版本，主要用于学术场合。他原本打算将其命名为“Freax”，但是 Torvalds 曾经用来分发原始代码的服务器管理员将 Torvalds 的名和 UNIX 组合，把他的目录命名为“Linux”。

6. 如何为 Linux 贡献力量

大多数 Linux 内核都是用 C 编程语言编写的，并使用了一些汇编语言和其他语言。如果你有兴趣为 Linux 内核本身编写代码，那么入门的好地方是 Kernel Newbies FAQ，它将解释你想熟悉的一些概念和过程。

但是 Linux 社区不仅仅是内核，它还需要除了程序员以外的许多其他人的贡献。每个发行版都包含成百上千个可以与之一起发行的程序，并且每个程序以及发行版本身都需要各种人员和技能来使其成功，包括：

- 测试人员，以确保一切都可以在不同的硬件和软件配置上正常运行，并在无法正常运行时报告错误。
- 设计人员创建随各种程序一起分发的用户界面和图形。
- 写作者创建文档、操作指南以及随软件分发的其他重要文本。
- 译者把用他们的母语写成的程序和文档翻译为其他语言，供世界各地的人们使用。
- 打包者把软件程序与其他所有部分捆绑在一起，以确保它们在不同发行版中都能正常运行。
- 传播者通常会传播有关 Linux 和开源的信息。
- 当然，开发人员也可以自己编写软件。

7. 如何开始使用 Linux

你可能已经在使用 Linux，但你并不知道，但是如果你想在家用计算机上安装 Linux 进行试用，最简单的方法是选择一种针对你的平台（例如笔记本计算机或平板计算机设备）的流行发行版，然后试一试。尽管有很多可用的发行版，但是大多数较早的知名发行版对于初学者来说都是不错的选择，因为它们拥有庞大的用户群体，如果你遇到困难或者无法解决问题时他们可以帮助你解决问题或遇到麻烦时回答问题。流行的发行版包括 Debian、Fedora、Mint 和 Ubuntu，但还有许多其他发行版本。

Exercises

[Ex. 1] Answer the following questions according to Text A.

1. What does an operating system act as? What is the purpose of an operating system?
2. What is an operating system concerned with?
3. How many functions does an operating system perform? What are they?
4. What are the advantages of batch operating system?
5. What are the disadvantages of time-sharing OS?
6. What is the major benefit of working with distributed operating system?
7. What is one more important aspect of network operating systems?
8. What are the disadvantages of network operating system?

9. What are hard real-time systems meant for?
10. What are the disadvantages of RTOS?

[Ex.2] Answer the following questions according to Text B.

1. What is Linux?
2. What is first, and perhaps most important about Linux?
3. Why is Linux incredibly customizable?
4. When and where was Unix developed? And by whom?
5. Why do companies and individuals choose Linux for their servers?
6. Why is contacting each individual author and getting them to agree to a new license virtually impossible?
7. When and by whom was Linux created? How came the name Linux?
8. What is most of the Linux kernel written in?
9. What do testers do?
10. Why are most of the older, well-known distributions good choices for beginners?

[Ex. 3] Translate the following terms or phrases from English into Chinese and vice versa.

1. distributed operating system	1. ______
2. interrupt signal	2. ______
3. multitasking system	3. ______
4. tightly coupled system	4. ______
5. distributed operating system	5. ______
6. *n.*配给，分配	6. ______
7. *adj.*自治的，有自主权的	7. ______
8. *n.*布局，构造；配置	8. ______
9. *n.*模块；组件	9. ______
10. *n.*可靠，可靠性	10. ______

[Ex. 4] Translate the following sentences into Chinese.

OS

An operating system (OS) is the program that, after being initially loaded into the computer by a boot program, manages all of the other application programs in a computer. The application programs make use of the operating system by making requests for services through a defined application program interface (API). In addition, users can interact directly with the operating system through a user interface such as a command line or a graphical user interface (GUI).

An operating system can perform the following services for applications:

- In a multitasking operating system, where multiple programs can be running at the same time, the OS determines which applications should be run in what order and how much time should be allowed for each application before giving another application a turn.
- It manages the sharing of internal memory among multiple applications.
- It handles input and output to and from attached hardware devices, such as hard disks, printers and ports.
- It sends messages to each application or interactive user (or to a system operator) about the status of operation and any errors that may have occurred.
- It can offload the management of batch jobs (for example, printing) so that the initiating application is freed from this work.
- On computers that can provide parallel processing, an operating system can manage how to divide the program so that it runs on more than one processor at a time.

All major computer platforms (hardware and software) require an operating system, and operating systems must be developed with different features to meet the specific needs of various form factors.

A mobile OS allows smartphones, tablet PCs and other mobile devices to run applications and programs.

An embedded operating system is specialized for use in the computers built into larger systems, such as cars, traffic lights, digital televisions, ATMs, airplane controls, point of sale (POS) terminals, digital cameras, GPS navigation systems, elevators, digital media receivers and smart meters.

A network operating system (NOS) is a computer operating system that is designed primarily to support workstation, personal computer, and, in some instances, older terminals that are connected on a local area network (LAN).

A real-time operating system (RTOS) is an operating system that guarantees a certain capability within a specified time constraint. For example, an operating system might be designed to ensure that a certain object was available for a robot on an assembly line.

[Ex. 5] Fill in the blanks with the words given below.

users	functions	different	commands	desktop
provide	system	run	clicking	operating

OS

The operating system (OS) is the most important program that runs on a computer. Every general-purpose computer must have an operating ___1___ to run other programs and applications. Computer operating systems perform basic tasks, such as recognizing input from the keyboard, sending output to the display screen, keeping track of files and directories on the storage drives, and controlling peripheral devices, such as printers.

For large systems, the operating system has even greater responsibilities and powers. It is like a traffic cop — it makes sure that different programs and ____2____ running at the same time do not interfere with each other. The operating system is also responsible for security, ensuring that unauthorized users do not access the system.

1. A Software Platform for Applications

Operating systems ____3____ a software platform on top of which other programs, called application programs, can run. The application programs must be written to run on top of a particular ____4____ system. Your choice of operating system, therefore, determines to a great extent the applications you can run.

2. Classification of Operating Systems

- Multi-user: Allows two or more users to ____5____ programs at the same time. Some operating systems permit hundreds or even thousands of concurrent users.
- Multiprocessing: Supports running a program on more than one CPU.
- Multitasking: Allows more than one program to run concurrently.
- Multithreading: Allows ____6____ parts of a single program to run concurrently.
- Real time: Responds to input instantly.

3. User Interaction with the OS

As a user, you normally interact with the operating system through a set of ____7____. The commands are accepted and executed by a part of the operating system called the command processor or command line interpreter. Graphical user interfaces allow you to enter commands by pointing and ____8____ at objects that appear on the screen.

4. Mobile Operating Systems

In the same way that a desktop OS controls your ____9____ or laptop computer, a mobile operating system is the software platform on top of which other programs can run on mobile devices, however, these systems are designed specifically to run on mobile devices such as mobile phones, smartphones, PDAs, tablet computers and other handhelds.

The mobile OS is responsible for determining the ____10____ and features available on your device, such as thumb wheel, keyboards, WAP, synchronization with applications, email, text messaging and more. The mobile OS will also determine which third-party applications (mobile apps) can be used on your device.

Online Resources

二维码	内　容
	计算机专业常用语法（3）：动词不定式

二维码	内　　容
	在线阅读（1）：UNIX （UNIX 操作系统）
	在线阅读（2）：Network Operating Systems （网络操作系统）

Unit 4

Data Structure and Algorithm

Text A

扫码听课文

Data Structure

Simply put, a data structure is a container that stores data in a specific layout. This "layout" allows a data structure to be efficient in some operations and inefficient in others. The following are the top eight data structures that you should know.

1. Arrays

An array is the simplest and most widely used data structure. Other data structures like stacks and queues are derived from arrays.

Here's an image of a simple array of size 4, containing elements (1, 2, 3 and 4) (see Figure 4-1).

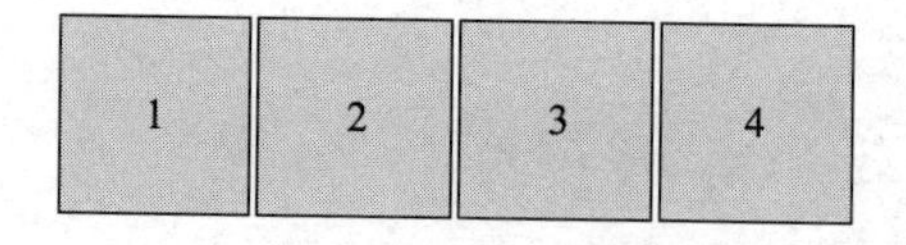

Figure 4-1 Array element

Each data element is assigned a positive numerical value called the index, which corresponds to the position of that item in the array. The majority of languages define the starting index of the array as 0.

The following are the two types of arrays:

- One-dimensional arrays (as shown above)
- Multi-dimensional arrays (arrays within arrays)

Basic operations of arrays:

- Insert — Inserts an element at given index.
- Get — Returns the element at given index.
- Delete — Deletes an element at given index.
- Size — Gets the total number of elements in array.

Commonly asked array interview questions:

- Find the second minimum element of an array.
- Find first non-repeating integers in an array.
- Merge two sorted arrays.
- Rearrange positive and negative values in an array.

2. Stacks

We are all familiar with the famous Undo option, which is present in almost every application. Ever wondered how it works? The idea: you store the previous states of your work (which are limited to a specific number) in the memory in such an order that the last one appears first. This can't be done just by using arrays. That is where the stack comes in handy.

A real-life example of stack could be a pile of books placed in a vertical order. In order to get the book that's somewhere in the middle, you will need to remove all the books placed on top of it. This is how the LIFO (last in first out) method works.

Here's an image of stack containing three data elements (1, 2 and 3), where 3 is at the top and will be removed first (see Figure 4-2).

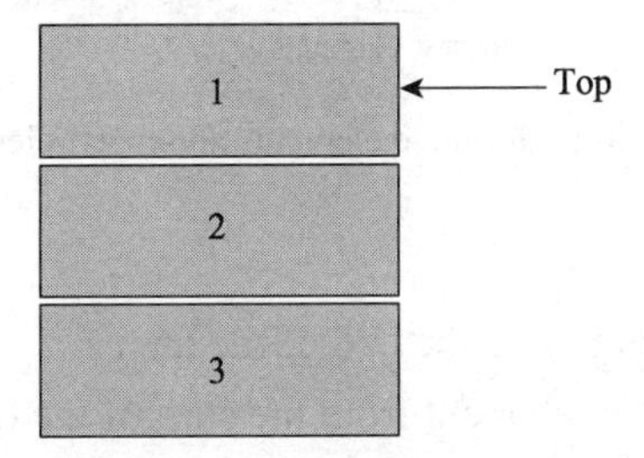

Figure 4-2 Stack containing three data elements

Basic operations of stack:

- Push — Inserts an element at the top.
- Pop — Returns the top element after removing from the stack.
- isEmpty — Returns true if the stack is empty.
- Top — Returns the top element without removing from the stack.

Commonly asked stack interview questions:

- Evaluate postfix expression using a stack.
- Sort values in a stack.
- Check balanced parentheses in an expression.

3. Queues

Similar to stack, queue is another linear data structure that stores the element in a sequential manner. The only significant difference between stack and queue is that instead of using the LIFO method, queue implements the FIFO method, which is short for first in first out.

A perfect real-life example of queue: a line of people waiting at a ticket booth. If a new person comes, they will join the line from the end, not from the start — and the person standing at the front will be the first to get the ticket and hence leave the line.

Here's an image of queue containing four data elements (1, 2, 3 and 4), where 1 is at the top and will be removed first (see Figure 4-3).

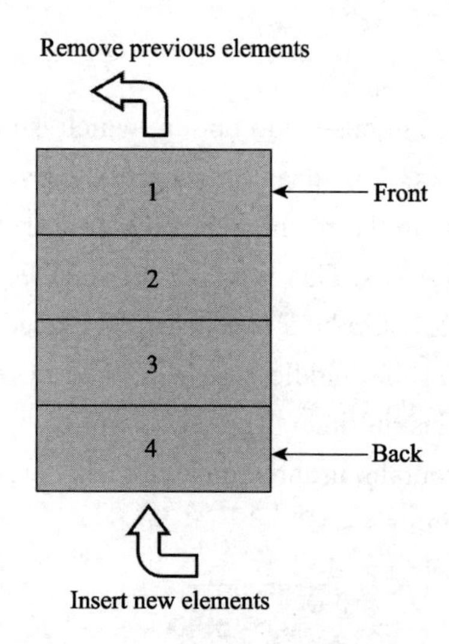

Figure 4-3 Remove elements and insert elements

Basic operations of Queue:

- Enqueue() — Inserts element to the end of the queue.
- Dequeue() — Removes an element from the start of the queue.
- isEmpty() — Returns true if queue is empty.
- Top() — Returns the first element of the queue.

Commonly asked queue interview questions:

- Implement stack using a queue.
- Reverse first *k* elements of a queue.
- Generate binary numbers from 1 to *n* using a queue.

4. Linked List

A linked list is another important linear data structure which might look similar to arrays at first but differs in memory allocation, internal structure and how basic operations of insertion and deletion are carried out.

A linked list is like a chain of nodes, where each node contains information like data and a pointer to the succeeding node in the chain. There's a head pointer, which points to the first element of the linked list, and if the list is empty then it simply points to null or nothing.

Linked lists are used to implement file systems, hash tables, and adjacency lists.

Here's a visual representation of the internal structure of a linked list (see Figure 4-4).

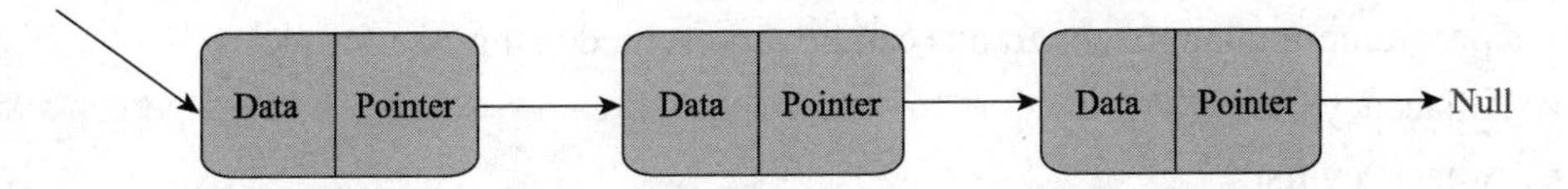

Figure 4-4 Internal structure of a linked list

Following are the types of linked lists:

- Singly linked list (Unidirectional)
- Doubly linked list (Bidirectional)

Basic operations of linked list:

- InsertAtEnd — Inserts given element at the end of the linked list.
- InsertAtHead — Inserts given element at the start/head of the linked list.
- Delete — Deletes given element from the linked list.
- DeleteAtHead — Deletes first element of the linked list.
- Search — Returns the given element from a linked list.
- isEmpty — Returns true if the linked list is empty.

Commonly asked linked list interview questions:

- Reverse a linked list.
- Detect loop in a linked list.
- Return *N*th node from the end in a linked list.
- Remove duplicates from a linked list.

5. Graphs

A graph is a set of nodes that are connected to each other in the form of a network. Nodes are also called vertices. A pair(x,y) is called an edge, which indicates that vertex x is connected to vertex y. An edge may contain weight/cost, showing how much cost is required to traverse from vertex x to y (see Figure 4-5).

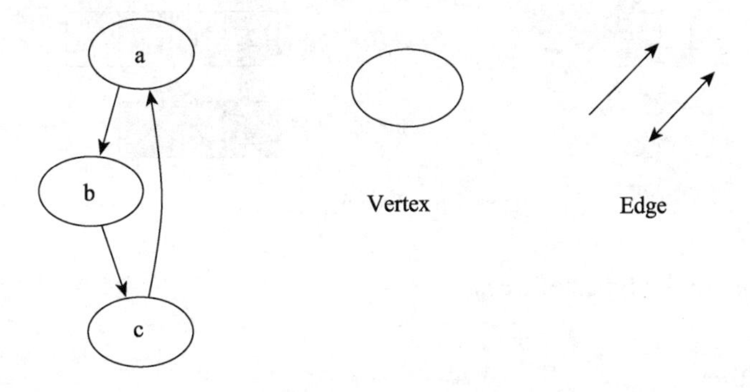

Figure 4-5 Types of graphs

Types of graphs:

- Undirected graph
- Directed graph

In a programming language, graphs can be represented using two forms:

- Adjacency matrix
- Adjacency list

Common graph traversing algorithms:

- Breadth first search
- Depth first search

Commonly asked graph interview questions:

- Implement breadth and depth first search.
- Check if a graph is a tree or not.
- Count number of edges in a graph.
- Find the shortest path between two vertices.

6. Trees

A tree is a hierarchical data structure consisting of vertices (nodes) and edges that connect them. Trees are similar to graphs, but the key point that differentiates a tree from the graph is that a cycle cannot exist in a tree.

Trees are extensively used in artificial intelligence and complex algorithms to provide an efficient storage mechanism for problem-solving.

Here's an image of a simple tree, and basic terminologies used in tree data structure (see Figure 4-6).

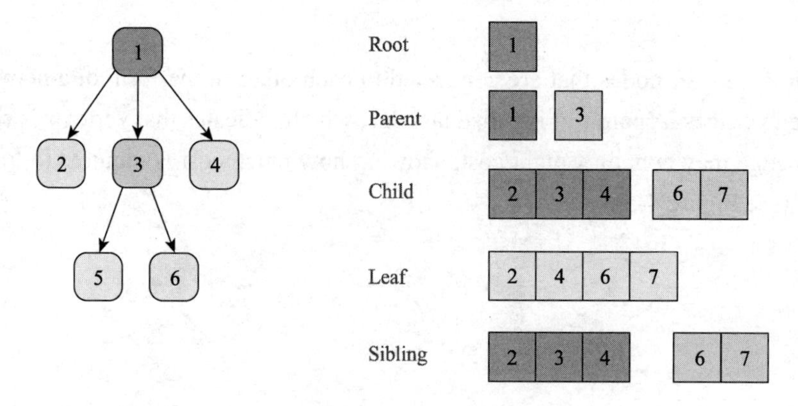

Figure 4-6 Tree data structure

The following are the types of trees:

- N-ary tree
- Balanced tree
- Binary tree
- Binary search tree
- AVL tree
- Red black tree

- 2–3 tree

Out of the above, binary tree and binary search tree are the most commonly used trees.

Commonly asked tree interview questions:

- Find the height of a binary tree.
- Find *k*th maximum value in a binary search tree.
- Find nodes at "*k*" distance from the root.
- Find ancestors of a given node in a binary tree.

7. Trie

Trie, which is also known as "prefix trees", is a tree-like data structure which proves to be quite efficient for solving problems related to strings. It provides fast retrieval, and is mostly used for searching words in a dictionary, providing auto suggestions in a search engine, and even for IP routing.

Here's an illustration of how the three words "top", "thus", and "their" are stored in trie (see Figure 4-7).

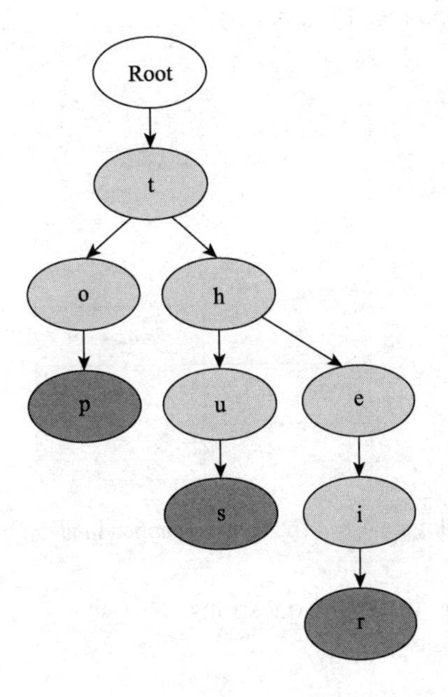

Figure 4-7 Stored in trie

The words are stored in the top to the bottom manner where nodes "p", "s" and "r" indicates the end of "top", "thus", and "their" respectively.

Commonly asked trie interview questions:

- Count total number of words in trie.
- Print all words stored in trie.
- Sort elements of an array using trie.

- Form words from a dictionary using trie.
- Build a T9 dictionary.

8. Hash Table

Hashing is a process used to uniquely identify objects and store each object at some pre-calculated unique index called its "key". So, the object is stored in the form of a "key-value" pair, and the collection of such items is called a "dictionary." Each object can be searched using that key. There are different data structures based on hashing, but the most commonly used data structure is the hash table.

Hash tables are generally implemented using arrays.

The performance of hashing data structure depends upon these three factors:

- Hash function
- Size of the hash table
- Collision handling method

Here's an illustration of how the hash is mapped in an array. The index of this array is calculated through a hash function (see Figure 4-8).

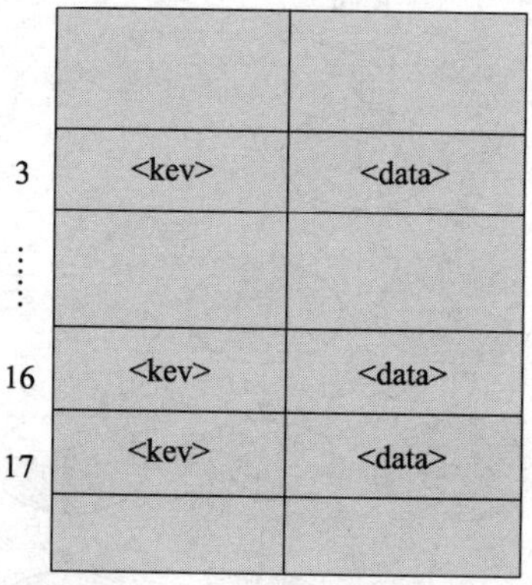

3	<kev>	<data>
⋮		
16	<kev>	<data>
17	<kev>	<data>

Figure 4-8 The hash is mapped in an array

Commonly asked hashing interview questions:

- Find symmetric pairs in an array.
- Trace complete path of a journey.
- Find if an array is a subset of another array.
- Check if given arrays are disjoint.

New Words

structure	[ˈstrʌktʃə]	*n.*结构，构造；体系
container	[kənˈteɪnə]	*n.*容器
inefficient	[ˌɪnɪˈfɪʃnt]	*adj.*无效率的，无能的

array	[ə'reɪ]	*n.*数组；队列，阵列
stack	[stæk]	*n.*堆栈
queue	[kjuː]	*n.*队列
element	['elɪmənt]	*n.*元素；要素
assign	[ə'saɪn]	*v.*赋值
positive	['pɒzətɪv]	*adj.*正的；积极的；确实的
index	['ɪndeks]	*n.*下标，标志；索引；指数
position	[pə'zɪʃn]	*n.*位置，方位；地位，职位；态度；状态 *vt.*安置；把……放在适当位置；给……定位
one-dimensional	[wʌn-dɪ'menʃənəl]	*adj.*一维的
multi-dimensional	[ˌmʌltɪ-daɪ'menʃənl]	*adj.*多维的
insert	[ɪn'sɜːt]	*vt.*插入
return	[rɪ'tɜːn]	*v.*返回，回来
delete	[dɪ'liːt]	*v.*删除
interview	['ɪntəvjuː]	*n.*面试；会谈
non-repeating	[nɒn-rɪ'piːtɪŋ]	*adj.*非重复的
integer	['ɪntɪdʒə]	*n.*整数
merge	[mɜːdʒ]	*v.*合并；（使）混合
rearrange	[ˌriːə'reɪndʒ]	*vt.*重排；重新布置
negative	['negətɪv]	*adj.*[数]负的
state	[steɪt]	*n.*状况，情况
push	[pʊʃ]	*v.*入栈
pop	[pɒp]	*v.*出栈
empty	['emptɪ]	*adj.*空的
evaluate	[ɪ'væljʊeɪt]	*vt.*评价；求……的值（或数）
expression	[ɪk'spreʃn]	*n.*表达式；表现，表示
parenthesis	[pə'renθəsɪs]	*n.*圆括号
sequential	[sɪ'kwenʃl]	*adj.*序列的，按次序的
reverse	[rɪ'vɜːs]	*v.*（使）反转；（使）颠倒；调换，交换
generate	['dʒenəreɪt]	*vt.*产生，造成
insertion	[ɪn'sɜːʃn]	*n.*插入
deletion	[dɪ'liːʃn]	*n.*删除
node	[nəʊd]	*n.*节点
pointer	['pɔɪntə]	*n.*指针

null	[nʌl]	*adj*.空的；零值的，等于零的
representation	[ˌreprɪzenˈteɪʃn]	*n*.表现；陈述
unidirectional	[ˌju:nɪdɪ'rekʃənəl]	*adj*.单向的
bidirectional	[ˌbɑɪdəˈrekʃənl]	*adj*.双向的
loop	[lu:p]	*n*.循环
graph	[græf]	*n*.图
vertex	[ˈvɜ:teks]	*n*.顶点
vertices	[ˈvɜ:tɪsi:z]	*n*.顶点（vertex 的名词复数）
edge	[edʒ]	*n*.边
indicate	[ˈɪndɪkeɪt]	*vt*.表明，标示，指示
traverse	[trəˈvɜ:s]	*n*.穿过；横贯
		vt.通过；横越，横贯
adjacency	[ə'dʒeɪsnsɪ]	*n*.邻接
matrix	[ˈmeɪtrɪks]	*n*.矩阵
root	[ru:t]	*n*.根
ancestor	[ˈænsestə]	*n*.祖先，祖宗
string	[strɪŋ]	*n*.字符串
retrieval	[rɪˈtri:vl]	*n*.检索；收回，挽回
route	[ru:t]	*n*.路由
		vt.按某路线发送
illustration	[ˌɪləˈstreɪʃn]	*n*.说明；例证；图解
respectively	[rɪˈspektɪvlɪ]	*adv*.各自地；各个地；分别地
sort	[sɔ:t]	*vt*.将……排顺序；把……分类
		n.排序；分类
hash	[hæʃ]	*vt*.散列
pre-calculated	[ˈpri:-ˈkælkjuleɪtɪd]	*adj*.预计算的
collection	[kəˈlekʃn]	*n*.收集，采集
factor	[ˈfæktə]	*n*.因素
map	[mæp]	*vt*.映射
symmetric	[sɪ'metrɪk]	*adj*.对称的；均衡的
subset	[ˈsʌbset]	*n*.子集
disjoint	[dɪs'dʒɔɪnt]	*adj*.不相交的，无交集的

Phrases

data structure	数据结构

positive numerical	正数
the majority of	大多数
one-dimensional array	一维数组
multi-dimensional array	多维数组
come in handy	派得上用场，迟早有用
a pile of	一摞，一堆
postfix expression	后缀表达式
linear data structure	线性数据结构
linked list	链表
memory allocation	内存分配
carry out	执行，完成
a chain of	一链，一串；一系列
head pointer	头指针，首指针
hash table	散列表，哈希表
adjacency list	邻接表
singly linked list	单链表，单向链表
doubly linked list	双链表，双向链表
a set of	一套，一组
undirected graph	无向图
directed graph	有向图
adjacency matrix	邻接矩阵
breadth first search	广度优先搜索
depth first search	深度优先搜索
hierarchical data structure	分层数据结构
artificial intelligence	人工智能
N-ary tree	N 叉树
balanced tree	均衡树
binary tree	二叉树
binary search tree	二叉搜索树，二叉查找树，二叉排序树
AVL tree	AVL 树，得名于它的发明者 G. M. Adelson-Velsky 和 E. M. Landis
Prefix Tree	字典树，前缀树
auto suggestion	自动建议
depend upon	根据，依据；依靠
collision handling method	碰撞处理方法

Abbreviations

LIFO (Last In First Out)	后进先出
FIFO (First In First Out)	先进先出
IP (Internet Protocol)	网际互连协议

Text A 参考译文

数 据 结 构

简而言之，数据结构是一个以特定布置存储数据的容器。这种“布置”使数据结构在某些操作中有效，而在另一些操作中效率低下。以下是你应该了解的 8 个数据结构。

1. 数组

数组是最简单、使用最广泛的数据结构。其他数据结构（如堆栈和队列）是从数组派生而来。

这是大小为 4 的简单数组的图像，其中包含元素（1、2、3 和 4）。

（图略）

每个数据元素都被分配一个称为下标的正数值，该数值对应于该项在数组中的位置。大多数语言将数组的起始下标定义为 0。

以下是两种类型的数组：

- 一维数组（如上所示）。
- 多维数组（数组中的数组）。

数组的基本操作：

- 插入：在给定下标处插入元素。
- 获取：返回给定下标处的元素。
- 删除：删除给定下标处的元素。
- 大小：获取数组中元素的总数。

数组面试常见问题：

- 查找数组的第二小的元素。
- 查找数组中的第一个非重复整数。
- 合并两个排序的数组。
- 在数组中重新排列正值和负值。

2. 堆栈

我们都熟悉著名的撤销选项，目前几乎所有的应用软件都有这一选项。有没有想过它是如何工作的？这个想法是：按照把最后一个放在最前面的顺序，将工作的先前状态（限于特定数量）存储在内存中。这不能仅通过使用数组来完成。此时堆栈就派上用场了。

现实生活中垂直顺序放置的一摞书可以是堆栈的示例。为了拿到中间的书籍，你需要拿开所有放在其上面的书籍。这就是 LIFO（后进先出）方法的工作方式。

这是包含三个数据元素（1、2 和 3）的堆栈图像，其中 3 在顶部，并且将首先移去。

（图略）

堆栈的基本操作：

- Push：在顶部插入一个元素。
- Pop：返回顶部元素并将其从堆栈中移除。
- IsEmpty：如果堆栈为空，则返回 true。
- Top：返回顶部元素，但不从堆栈中移除。

堆栈面试常见问题：

- 使用堆栈计算后缀表达式。
- 对堆栈中的值进行排序。
- 检查表达式中的括号是否配对。

3. 队列

类似于堆栈，队列是另一种线性数据结构，它以顺序方式存储元素。堆栈和队列之间的唯一显著区别是，队列使用 FIFO 方法代替了 LIFO 方法，该方法是先进先出的缩写。

一个真实的队列示例：人们在售票亭排队等待。如果有新人来，将排在队尾而不是队首——站在最前面的人将第一个获得票并离开。

这是一张包含 4 个数据元素（1、2、3 和 4）的队列图像，其中 1 位于顶部，并且将首先移去。

（图略）

队列的基本操作：

- Enqueue()：将元素插入队列的末尾。
- Dequeue()：从队列的开头删除一个元素。
- isEmpty()：如果队列为空，则返回 true。
- Top()：返回队列的第一个元素。

队列面试常见问题：

- 使用队列实现堆栈。

- 反转队列的前 k 个元素。
- 使用队列生成从 1 到 n 的二进制数。

4. 链表

链表是另一种重要的线性数据结构，乍一看可能与数组相似，但在内存分配、内部结构以及插入和删除的基本操作方式上有所不同。

链表就像一个节点链，其中每个节点都包含信息（例如数据）和指向链中后续节点的指针。有一个头指针，它指向链表的第一个元素，如果列表为空，则仅指向 null 或不指向任何内容。

链表用于实现文件系统、哈希表和邻接表。

这是链表内部结构的直观表示。

（图略）

以下是链表的类型：

- 单链表（单向）。
- 双链表（双向）。

链表的基本操作：

- InsertAtEnd：在链表的末尾插入给定的元素。
- InsertAtHead：在链接列表的开头插入给定的元素。
- Delete：从链表中删除给定的元素。
- DeleteAtHead：删除链表的第一个元素。
- Search：从链表中返回给定的元素。
- IsEmpty：如果链接列表为空，则返回 true。

常见的链表面试问题：

- 反向链表。
- 在链表中检测循环。
- 从链表的末尾返回第 N 个节点。
- 删除链表中的重复项。

5. 图

图是一组以网络形式相互连接的节点。节点也称为顶点。一对(x,y)称为边，它表示顶点 x 连接到顶点 y 的边。一条边可能包含重量/成本，表明从顶点 x 到 y 遍历需要多少成本。

（图略）

图的类型：

- 无向图。

- 有向图。

在编程语言中，图形可以使用两种形式表示：

- 邻接矩阵。
- 邻接表。

常见的图遍历算法：

- 广度优先搜索。
- 深度优先搜索。

图面试常见问题：

- 进行广度和深度优先搜索。
- 检查图形是否为树。
- 计算图中的边数。
- 查找两个顶点之间的最短路径。

6. 树

树是由顶点（节点）和连接它们的边组成的分层数据结构。树类似于图，但是区别图与树的关键是树中不能存在循环。

树被广泛用于人工智能和复杂算法中，以提供有效的存储机制来解决问题。

这是一棵简单的树的图像，以及树数据结构中使用的基本术语。

（图略）

以下是树的类型：

- N 叉树。
- 平衡树。
- 二叉树。
- 二叉搜索树。
- AVL 树。
- 红黑树。
- 2-3 树。

其中，二叉树和二叉搜索树是最常用的树。

树面试常见问题：

- 查找二叉树的高度。
- 在二叉搜索树中找到第 k 个最大值。
- 查找距离根节点为“k”的节点。
- 在二叉树中查找给定节点的先辈。

7. 字典树

字典树也称为“前缀树”，是一种类似树的数据结构，被证明对于解决与字符串有关的问题非常有效。它提供了快速的检索功能，主要用于在字典中搜索单词，在搜索引擎中提供自动建议，甚至用于IP路由。

下面说明了如何在字典树中存储“top”“thus”和“their”这三个词。

（图略）

单词以从上到下的方式存储，其中节点“p”“s”和“r”分别表示“top”“thus”和“their”的结尾。

字典树面试常见问题：

- 计算字典树中的单词总数。
- 打印所有存储在字典树中的单词。
- 使用字典树对数组的元素进行排序。
- 使用字典树从字典中形成单词。
- 构建T9词典。

8. 哈希表

散列是一种过程，专门用于标识对象并将每个对象存储在某个预先计算的唯一索引（称为“键”）中。因此，对象以“键值”对的形式存储，而这些项目的集合被称为“字典”。可以使用该键搜索每个对象。虽然基于散列的数据结构有多种，但是最常用的数据结构是哈希表。

哈希表通常使用数组来实现。

散列数据结构的性能取决于以下三个因素。

- 哈希函数。
- 哈希表的大小。
- 碰撞处理方法。

这里说明了哈希如何映射到数组中。该数组的下标是通过哈希函数计算的。

（图略）

散列面试常见问题：

- 在数组中查找对称对。
- 跟踪旅程的完整路径。
- 查找一个数组是另一个数组的子集。
- 检查给定的数组是否不相交。

Text B

Algorithm

扫码听课文

An algorithm is a set of steps of operations to solve the problem of performing calculation, data processing, and automated reasoning tasks. An algorithm is an efficient method that can be expressed within finite amount of time and space.

An algorithm is the best way to represent the solution of a particular problem in a very simple and efficient way. If we have an algorithm for a specific problem, we can implement it in any programming language, meaning that the algorithm is independent from any programming languages.

1. Algorithm Design

The important aspects of algorithm design include creating an efficient algorithm to solve a problem in an efficient way using minimum time and space.

To solve a problem, different approaches can be followed. Some of them can be efficient with respect to time consumption, whereas other approaches may be memory efficient. However, one has to keep in mind that both time consumption and memory usage cannot be optimized simultaneously. If we require an algorithm to run in lesser time, we have to invest in more memory and if we require an algorithm to run with lesser memory, we need to have more time.

2. Problem Development Steps

The following steps are involved in solving computational problems.

- Problem definition
- Development of a model
- Specification of an algorithm
- Designing an algorithm
- Checking the correctness of an algorithm
- Analysis of an algorithm
- Implementation of an algorithm
- Program testing
- Documentation

3. Characteristics of Algorithms

The main characteristics of algorithms are as follows:

- Algorithms must have a unique name.
- Algorithms should have explicitly defined set of inputs and outputs.

- Algorithms are well-ordered with unambiguous operations.
- Algorithms halt in a finite amount of time. Algorithms should not run for infinity, i.e., an algorithm must end at some point.

4. Pseudocode

Pseudocode gives a high-level description of an algorithm without the ambiguity associated with plain text but also without the need to know the syntax of a particular programming language.

The running time can be estimated in a more general manner by using pseudocode to represent the algorithm as a set of fundamental operations.

Difference between algorithm and pseudocode:

An algorithm is a formal definition with some specific characteristics that describes a process, which could be executed by a Turing-complete computer machine to perform a specific task. Generally, the word "algorithm" can be used to describe any high level task in computer science.

On the other hand, pseudocode is an informal and (often rudimentary) human readable description of an algorithm leaving many granular details of it. Writing a pseudocode has no restriction of styles and its only objective is to describe the high level steps of algorithm in a much realistic manner in natural language.

The following is an example of an algorithm for Insertion-Sort.

```
Algorithm: Insertion-Sort
Input: A list L of integers of length n
Output: A sorted list L1 containing those integers present in L
Step 1: Keep a sorted list L1 which starts off empty
Step 2: Perform Step 3 for each element in the original list L
Step 3: Insert it into the correct position in the sorted list L1
Step 4: Return the sorted list
Step 5: Stop
```

Here is a pseudocode which describes how the high-level abstract process mentioned above in the algorithm Insertion-Sort could be described in a more realistic way.

```
for i <- 1 to length(A)
  x <- A[i]
  j <- i
  while j > 0 and A[j-1] > x
    A[j] <- A[j-1]
    j <- j - 1
  A[j] <- x
```

In this passage, algorithms will be presented in the form of pseudocode, which is similar in many respects to C, C++, Java, Python, and other programming languages.

In theoretical analysis of algorithms, it is common to estimate their complexity in the asymptotic sense, that is to estimate the complexity function for arbitrarily large input. The term "analysis of algorithms" was coined by Donald Knuth.

Algorithm analysis is an important part of computational complexity theory, which provides theoretical estimation for the required resources of an algorithm to solve a specific computational problem. Most algorithms are designed to work with inputs of arbitrary length. Analysis of algorithms is the determination of the amount of time and space resources required to execute it.

Usually, the efficiency or running time of an algorithm is stated as a function relating the input length to the number of steps, known as time complexity, or volume of memory, known as space complexity.

5. The Need for Analysis

In this section, we will discuss the need for analysis of algorithms and how to choose a better algorithm for a particular problem as one computational problem can be solved by different algorithms.

By considering an algorithm for a specific problem, we can begin to develop pattern recognition so that similar types of problems can be solved with the help of this algorithm.

Algorithms are often quite different from one another, though the objective of these algorithms are the same. For example, we know that a set of numbers can be sorted using different algorithms. The number of comparisons performed by one algorithm may vary with others for the same input. Hence, time complexity of those algorithms may differ. At the same time, we need to calculate the memory space required by each algorithm.

Analysis of algorithm is the process of analyzing the problem-solving capability of the algorithm in terms of the time and size required (the size of memory for storage while implementation). However, the main concern of analysis of algorithms is the required time or performance. Generally, we perform the following types of analysis:

- Worst case—The maximum number of steps taken on any instance of size *a*.
- Best case—The minimum number of steps taken on any instance of size *a*.
- Average case—An average number of steps taken on any instance of size *a*.
- Amortized—A sequence of operations applied to the input of size *a* averaged over time.

To solve a problem, we need to consider time as well as space complexity as the program may run on a system where memory is limited but adequate space is available or may be vice-versa. In this context, if we compare bubble sort and merge sort, bubble sort does not require additional memory, but merge sort requires additional space. Though time complexity of bubble sort is higher compared to merge sort, we may need to apply bubble sort if the program needs to run in an environment, where memory is very limited.

New Words

algorithm ['ælgərɪðəm] *n.*算法

express	[ɪkˈspres]	*vt.*表达，表示
finite	[ˈfaɪnaɪt]	*adj.*有限的，有穷的 *n.*有限性
minimum	[ˈmɪnɪməm]	*adj.*最小的，最少的 *n.*最小量
approach	[əˈprəʊtʃ]	*n.*方法；途径
correctness	[kə'rektnəs]	*n.*正确性
documentation	[ˌdɒkjʊmenˈteɪʃn]	*n.*文档
unique	[jʊˈni:k]	*adj.*唯一的，仅有的
explicitly	[ɪk'splɪsɪtlɪ]	*adv.*明白地，明确地
unambiguous	[ˌʌnæmˈbɪgjʊəs]	*adj.*清楚的，明白的
infinity	[ɪnˈfɪnətɪ]	*n.*无穷，无限
pseudocode	['sju:dəʊˌkəʊd]	*n.*伪代码
description	[dɪˈskrɪpʃn]	*n.*描述，形容
ambiguity	[ˌæmbɪˈgju:ətɪ]	*n.*含糊，意义不明确
estimate	[ˈestɪmeɪt]	*vt.*估计，估算
Turing-complete	[ˈtʊəriŋ-kəmˈpli:t]	*adj.*图灵完备的
informal	[ɪnˈfɔ:ml]	*adj.*非正式的
rudimentary	[ˌru:dɪˈmentərɪ]	*adj.*基本的，初步的
readable	[ˈri:dəbl]	*adj.*可读的；易懂的
restriction	[rɪˈstrɪkʃn]	*n.*限制，限定；拘束
realistic	[ˌri:əˈlɪstɪk]	*adj.*现实的
abstract	[ˈæbstrækt]	*adj.*抽象的 *n.*抽象概念；摘要 *vt.*提取，抽象
theoretical	[ˌθɪəˈretɪkl]	*adj.*理论的；推想的，假设的
asymptotic	[æsɪmp'tɒtɪk]	*adj.*渐近的
arbitrarily	['ɑ:bɪtrərɪlɪ]	*adv.*任意地
discuss	[dɪˈskʌs]	*vt.*讨论；论述，详述
amortize	[əˈmɔ:taɪz]	*vt.*平摊
adequate	[ˈædɪkwɪt]	*adj.*足够的；适当的

Phrases

data processing	数据处理
automated reasoning	自动推理

be independent from ...	与……无关的
time consumption	时间消耗，时间开销
invest in ...	在……投入，在……上投资
at some point	在某一时刻
insertion sort	插入排序
in the sense...	就……意义而言
computational complexity theory	计算复杂性理论，计算复杂度理论
be stated as ...	可表示为……
time complexity	时间复杂性
space complexity	空间复杂性
pattern recognition	模式识别
worst case	最坏情况
best case	最好情况
average case	平均情况
bubble sort	冒泡排序
merge sort	归并排序

Text B 参考译文

算　法

算法是一组操作步骤，用于解决执行计算、数据处理和自动推理任务的问题。算法是一种可以在有限的时间和空间内表达的有效方法。

算法是以非常简单和有效的方式表示特定问题的解决方案的最佳方法。如果我们有针对特定问题的算法，则可以使用任何编程语言来实现它，这意味着算法独立于任何编程语言。

1. 算法设计

算法设计的重要方面包括创建有效的算法以使用最少的时间和空间并以有效的方式解决问题。

为了解决问题，可以采用不同的方法。其中一些方法可能在时间消耗方面是有效的，而其他方法可能在内存使用方面是有效的。但是，必须记住，一个算法不能同时在时间消耗和内存使用方面都是优化的。如果我们需要一种算法在更短的时间内运行，那就必须投入更多的内存；如果我们需要一种算法在更少的内存中运行，那就需要更多的时间。

2. 问题开发步骤

解决计算问题涉及以下步骤。

- 问题定义。
- 模型开发。
- 算法说明。
- 设计算法。
- 检查算法的正确性。
- 算法分析。
- 算法的实现。
- 程序测试。
- 建立文档。

3. 算法特点

算法的主要特点如下：

- 算法必须具有唯一的名称。
- 算法应具有明确定义的一组输入和输出。
- 算法秩序井然、操作明确。
- 算法在有限的时间内终止。算法不应无限运行，即算法必须在某个时刻结束。

4. 伪代码

伪代码可以对算法进行高级描述，没有与纯文本相关联的歧义，也无须了解特定编程语言的语法。

通过使用伪代码将算法表示为一组基本运算，可以以一种更通用的方式估算运行时间。

算法与伪代码之间的区别：

算法是一种正式的定义，具有描述过程的某些特定特征，可以由图灵完备的计算机执行以执行特定任务。通常，“算法”一词可用于描述计算机科学中的任何高级任务。

另外，伪代码是算法的非正式且（通常是基本的）人类可读的描述，其中保留了许多细节。伪代码编写不受样式的限制，其唯一目的是用自然语言以更加现实的方式描述算法的高级步骤。

以下是插入-排序算法的示例。

算法：插入-排序。

输入：长度为 n 的整数的列表 L。

输出：包含 L 中存在的那些整数的排序列表 L1。

步骤 1：保留空的排序列表 L1。

步骤 2：对原始列表 L 中的每个元素执行步骤 3。

步骤 3：将其插入排序列表 L1 中的正确位置。

步骤 4：返回排序后的列表。

步骤 5：停止。

以下是一个伪代码，描述了如何以更实际的方式描述上述插入-排序算法中提到的高级抽象过程。

```
for i <- 1 to length(A)
  x <- A[i]
  j <- i
  while j > 0 and A[j-1] > x
    A[j] <- A[j-1]
    j <- j - 1
  A[j] <- x
```

在本文中，算法将以伪代码的形式呈现，这在许多方面与 C、C++、Java、Python 和其他编程语言相似。

在算法的理论分析中，通常在渐近意义上估计其复杂度，即估计任意大输入的复杂度函数。“算法分析”一词是由 Donald Knuth 创造的。

算法分析是计算复杂性理论的重要组成部分，它为解决特定计算问题所需的算法资源提供了理论上的估计。大多数算法被设计为可使用任意长度的输入。算法分析是为了确定执行算法所需的时间和空间资源。

通常，算法的效率或运行时间表示为将输入长度与步数相关联的函数（称为时间复杂度），或者表示为存储空间量（称为空间复杂度）。

5. 分析需求

在本节中，我们将讨论对算法进行分析的需求以及如何针对特定问题选择更好的算法，因为一个计算问题可以通过不同的算法来解决。

通过考虑针对特定问题的算法，我们可以开始开发模式识别，以便可以借助该算法解决类似类型的问题。

尽管这些算法的目标是相同的，但它们往往大相径庭。例如，我们知道可以使用不同的算法对一组数字进行排序。对于相同的输入，一种算法执行的比较次数可能会与其他算法有所不同。因此，这些算法的时间复杂度可能不同。同时，我们需要计算每种算法所需的存储空间。

算法分析是根据所需的时间和大小（实现时用于存储的内存大小）分析算法解决问题能力的过程。但是，算法分析的主要问题是所需的时间或性能。通常，我们执行以下类型的分析。

- 最坏情况：在任何大小为 a 的实例上执行的最大步骤数。

- 最佳情况：在任何大小为 a 的实例上执行的最小步骤数。
- 平均情况：在大小为 a 的任何实例上平均执行的步骤数。
- 平摊：应用于大小为 a 的输入与时间平均的一系列操作。

为了解决问题，我们需要考虑时间及空间复杂度，因为程序可能在内存有限但有足够可用空间的系统上运行，或者反之亦然。在这种情况下，如果我们比较冒泡排序和归并排序，冒泡排序不需要额外的内存空间，但是归并排序需要额外的内存空间。尽管冒泡排序的时间复杂度比归并排序要高，但是如果程序需要在内存非常有限的环境中运行，我们可能需要应用冒泡排序。

Exercises

[Ex. 1] Answer the following questions according to Text A.

1. What is a data structure?
2. What are the two types of arrays mentioned in the passage? What are the basic operations of arrays?
3. What could a real-life example of stack be? What are the basic operations of stack?
4. What is the only significant difference between stack and queue?
5. What are the commonly asked queue interview questions?
6. What is a link? What are the types of linked lists?
7. What are the types of graphs? How many forms can graphs be represented in a programming language?
8. What is a tree? What are the types of trees?
9. What is trie?
10. What is hashing? What does the performance of hashing data structure depend upon?

[Ex.2] Fill in the following blanks according to Text B.

1. An algorithm is a set of steps of operations to solve the problem of ____________, ____________, and ____________.
2. The important aspects of algorithm design include ____________ to solve a problem in an efficient way using ____________ and ____________.
3. The main characteristics of algorithms are as follows:

- Algorithms must have ____________.
- Algorithms should have explicitly defined set of ____________.
- Algorithms are well-ordered with ____________.
- Algorithms halt in a finite amount of time. Algorithms should not run for infinity, i.e., an algorithm must end ____________.

4. Pseudocode gives a high-level description of ________________ without the ambiguity ________________ but also without the need to ________________ of a particular programming language.
5. In theoretical analysis of algorithms, it is common to ________________ in the asymptotic sense, that is to estimate ________________ for ________________.
6. Algorithm analysis is an important part of ________________, which provides theoretical estimation for ________________ of an algorithm to solve ________________.
7. Usually, the efficiency or running time of an algorithm is stated as a function relating ________________ to ________________, known as time complexity, or volume of memory, known as ________________.
8. Analysis of algorithm is the process of analyzing ________________ of the algorithm in terms of ________________
9. Generally, we perform the following types of analysis: ________________, ________________, ________________ and ________________.
10. To solve a problem, we need to consider ____________ as well as ____________ as the program may run on a system where ____________ is limited but adequate space is ____________ or may be vice-versa.

[Ex. 3] Translate the following terms or phrases from English into Chinese and vice versa.

1. artificial intelligence	1.
2. binary search tree	2.
3. breadth first search	3.
4. doubly linked list	4.
5. hash table	5.
6. *n*.邻接	6.
7. *n*.数组；队列，阵列	7.
8. *adj*.双向的	8.
9. *n*.元素；要素	9.
10. *n*.表达式；表现，表示	10.

[Ex. 4] Translate the following sentences into Chinese.

Algorithms

Algorithm is a step-by-step procedure, which defines a set of instructions to be executed in a certain order to get the desired output. Algorithms are generally created independent of underlying languages, i.e. an algorithm can be implemented in more than one programming language.

From the data structure point of view, following are some important categories of algorithms:

- Search: Algorithm to search an item in a data structure.
- Sort: Algorithm to sort items in a certain order.
- Insert: Algorithm to insert item in a data structure.
- Update: Algorithm to update an existing item in a data structure.
- Delete: Algorithm to delete an existing item from a data structure.

1. Characteristics of an Algorithm

Not all procedures can be called an algorithm. An algorithm should have the following characteristics:

- Unambiguous: Algorithm should be clear and unambiguous. Each of its steps (or phases), and their inputs/outputs should be clear and must lead to only one meaning.
- Input: An algorithm should have 0 or more well-defined inputs.
- Output: An algorithm should have 1 or more well-defined outputs, and should match the desired output.
- Finiteness: Algorithms must terminate after a finite number of steps.
- Feasibility: Should be feasible with the available resources.
- Independent: An algorithm should have step-by-step directions, which should be independent of any programming code.

2. How to Write an Algorithm

There are no well-defined standards for writing algorithms. Rather, it is problem and resource dependent. Algorithms are never written to support a particular programming code.

As we know that all programming languages share basic code constructs like loops (do, for, while), flow-control (if-else), etc. These common constructs can be used to write an algorithm.

We write algorithms in a step-by-step manner, but it is not always the case. Algorithm writing is a process and is executed after the problem domain is well-defined. That is, we should know the problem domain, for which we are designing a solution.

[Ex. 5] Fill in the blanks with the words given below.

nodes	multiple	connected	array	tree
elements	structure	linked	level	linear

Key Differences Between Linear and Non-linear Data Structure

(1) In the linear data structure, the data is organized in a linear order in which elements are ___1___ one after the other. As against, in the non-linear data structure the data ___2___ are not stored in a sequential manner rather the elements are hierarchically related.

(2) The traversing of data in the ___3___ data structure is easy as it can make all the data

elements to be traversed in one go, but at a time only one element is directly reachable. On the contrary, in the non-linear data structure, the ___4___ are not visited sequentially and cannot be traversed in one go.

(3) Data elements are adjacently attached in the linear data ___5___, which means only two elements can be linked to two other elements while this is not the case in the non-linear data structure where one data element can be ___6___ to numerous other elements.

(4) The linear data structures are easily implemented relative to the non-linear data structure.

(5) A single ___7___ of elements is incorporated in the linear data structure. Conversely, non-linear data structure involves ___8___ levels.

(6) Examples of the linear data structure are ___9___, queue, stack, linked list, etc. In contrast, ___10___ and graph are the examples of the non-linear data structure.

(7) The memory is utilized efficiently in the non-linear data structure where linear data structure tends to waste the memory.

Online Resources

二维码	内　容
	计算机专业常用语法（4）：现在分词
	在线阅读（1）：Key Features of an Algorithm （算法的关键特征）
	在线阅读（2）：Machine Learning Algorithms and Their Applications （机器学习算法及其应用）

Unit 5

Database and Data Warehousing

Text A

Basic Concepts of Database

1. Database

A database is a collection of information that is organized so that it can easily be accessed, managed and updated. Databases can be classified according to types of content: bibliographic, full-text, numeric and images.

In computing, databases are sometimes classified according to their organizational approach. The most prevalent approach is the relational database, a tabular database in which data is defined so that it can be reorganized and accessed in a number of different ways. A distributed database is one that can be dispersed or replicated among different points in a network. An object-oriented programming database is one that is congruent with the data defined in object classes and subclasses.

Computer databases typically contain aggregations of data records or files, such as sales transactions, product catalogs and inventories, and customer profiles. Typically, a database manager provides users the capabilities of controlling read/write access, specifying report generation and analyzing usage. Databases and database managers are prevalent in large mainframe systems, but are also present in smaller distributed workstation and mid-range systems such as the AS/400 and on personal computers.

2. Relational Database

A relational database is a collection of data items organized as a set of formally-described tables from which data can be accessed or reassembled in many different ways without having to reorganize the database tables. The relational database was invented by E. F. Codd at IBM in 1970.

The standard user and application program interface to a relational database is the structured query language (SQL). SQL statements are used both for interactive queries for information from a relational database and for gathering data for reports.

In addition to being relatively easy to create and access, a relational database has the important advantage of being easy to extend. After the original database creation, a new data category can be added without requiring that all existing applications be modified.

A relational database is a set of tables containing data fitted into predefined categories. Each table (which is sometimes called a relation) contains one or more data categories in columns. Each row contains a unique instance of data for the categories defined by the columns. For example, a typical business order entry database would include a table that described a customer with columns for name, address, phone number, and so forth. Another table would describe an order: product, customer, date, sales price, and so forth. A user of the database could obtain a view of the database that fitted the user's needs. For example, a branch office manager might like a view or report on all customers that had bought products after a certain date. A financial services manager in the same company could, from the same tables, obtain a report on accounts that needed to be paid.

When creating a relational database, you can define the domain of possible values in a data column and further constraints that may apply to that data value. For example, a domain of possible customers could allow up to ten possible customer names but be constrained in one table to allowing only three of these customer names to be specifiable.

The definition of a relational database results in a table of metadata or formal descriptions of the tables, columns, domains, and constraints.

3. SQL

SQL (structured query language) is a standard language for making interactive queries from a database and updating a database such as IBM's DB2, Microsoft's Access and database products from Oracle, Sybase and Computer Associates. Although SQL is both an ANSI and an ISO standard, many database products support SQL with proprietary extensions to the standard language. Queries take the form of a command language that lets you select, insert, update, find out the location of data and so forth. There is also a programming interface.

4. Database Management System

A database management system (DBMS), sometimes just called a database manager, is a program that lets one or more computer users create and access data in a database. The DBMS manages user requests (and requests from other programs) so that users and other programs are free from having to understand where the data is physically located on storage media and, in a multi-user system, which else may also be accessing the data. In handling user requests, the DBMS ensures the integrity of the data (that is, making sure it continues to be accessible and is consistently organized as intended) and security (making sure only those with access privileges can access the data). The most typical DBMS is a relational database management system (RDBMS). A standard user and program interface is the structured query language (SQL). A newer kind of DBMS is the object-oriented database management system (ODBMS).

A DBMS can be thought of as a file manager that manages data in databases rather than files in file systems. In IBM's mainframe operating systems, the non-relational data managers were (and are, because these legacy application systems are still used) known as access methods.

A DBMS is usually an inherent part of a database product. On PCs, Microsoft's Access is a popular example of a single- or small-group user DBMS. Microsoft's SQL Server is an example of a DBMS that serves database requests from multiple (client) users. Other popular DBMSs (these are all RDBMSs, by the way) are IBM's DB2, Oracle's line of database management products, and Sybase's products.

IBM's information management system (IMS) was one of the first DBMSs. A DBMS may be used by or combined with transaction managers, such as IBM's customer information control system (CICS).

5. Distributed Database

A distributed database is a database in which portions of the database are stored on multiple computers within a network. Users have access to the portion of the database at their location so that they can access the data relevant to their tasks without interfering with the work of others.

6. DDBMS

A DDBMS (distributed database management system) is a centralized application that manages a distributed database as if it were all stored on the same computer. The DDBMS synchronizes all the data periodically, and in cases where multiple users must access the same data, ensures that updates and deletes performed on the data at one location will be automatically reflected in the data stored elsewhere.

7. Field

In a database table, a field is a data structure for a single piece of data. Fields are organized into records, which contain all the information within the table relevant to a specific entity. For example, in a table called customer contact information, telephone number would likely be a field in a row that would also contain other fields such as street address and city. The records make up the table rows and the fields make up the columns.

8. Record

In a database, a record (sometimes called a row) is a group of fields within a table that are relevant to a specific entity. For example, in a table called customer contact information, a row would likely contain fields such as: ID number, name, street address, city, telephone number and so on.

9. Table

In a relational database, a table (sometimes called a file) organizes the information about a single topic into rows and columns. For example, a database for a business would typically contain a table for customer information, which would store customers' account numbers, addresses, phone numbers, and so on as a series of columns. Each single piece of data (such as the account number) is a field in the table. A column consists of all the entries in a single field, such as the telephone numbers of all the customers. Fields, in turn, are organized as records, which are complete sets of information (such as the set of information about a particular customer), each of which comprises a row. The process of normalization determines how data will be most effectively organized into tables.

New Words

organize	[ˈɔ:gənaɪz]	*v.*组织
classify	[ˈklæsɪfaɪ]	*vt.*分类，分等
bibliographic	[bɪblɪə'græfɪk]	*adj.*书目的，目录的
approach	[əˈprəʊtʃ]	*n.*方法，步骤，途径，通路
tabular	[ˈtæbjʊlə]	*adj.*制成表的，扁平的，表格式的，平坦的
		*vi.*列表，排成表格式
disperse	[dɪˈspɜ:s]	*v.*(使)分散，(使)散开
congruent	[ˈkɒŋgrʊənt]	*adj.*（与 with 连用）一致的，适合的
aggregation	[ˌægrɪ'geɪʃn]	*n.*集合，集合体，聚合
catalog	[ˈkætəlɒg]	*n.*目录，目录册
		*v.*编目录
capability	[ˌkeɪpəˈbɪlɪtɪ]	*n.*(实际)能力，性能，容量
prevalent	[ˈprevələnt]	*adj.*普遍的，流行的
set	[set]	*n.*集合，集
reorganize	[rɪˈɔ:gənaɪz]	*v.*改组，再编制，改造
query	[ˈkwɪərɪ]	*v.*询问，查询
view	[vju:]	*n.*视图
domain	[dəˈmeɪn]	*n.*域，范围
specifiable	['spesɪfaɪəbl]	*adj.*能指定的；能详细说明的；能列举的
Oracle	[ˈɔrəkl]	*n.*美国甲骨文公司，主要生产数据库产品
ensure	[ɪn'ʃʊə]	*v.*确保
privilege	[ˈprɪvəlɪdʒ]	*n.*特权
inherent	[ɪnˈhɪərənt]	*adj.*固有的，内在的
centralize	[ˈsentrəlaɪz]	*vt.*集聚，集中

synchronize	[ˈsɪŋkrənaɪz]	*v*.同步
periodically	[ˌpɪərɪ'ɒdɪklɪ]	*adv*.周期性地，定时性地
automatically	[ˌɔ:tə'mætɪklɪ]	*adv*.自动地
reflect	[rɪˈflekt]	*v*.反射，反映，表现
field	[fi:ld]	*n*.域；字段
topic	[ˈtɒpɪk]	*n*.主题，题目
series	[ˈsɪəri:z]	*n*.连续，系列
complete	[kəmˈpli:t]	*adj*.完备的，完全的，完成的

Phrases

tabular database	表格数据库
distributed database	分布式数据库
customer profile	客户简介
find out	找出；发现

Abbreviations

SQL (Structured Query Language)	结构化查询语言
IBM (International Business Machines Corporation)	国际商用机器公司
ANSI (American National Standards Institute)	美国国家标准协会
ISO (International Organization for Standardization)	国际标准化组织
DBMS (Database Management System)	数据库管理系统
RDBMS (Relational Database Management System)	关系型数据库管理系统
ODBMS (Object-oriented Database Management System)	面向对象的数据库管理系统
IMS (Information Management System)	信息管理系统
CICS (Customer Information Control System)	客户信息管理系统
DDBMS (Distributed Database Management System)	分布式数据库管理系统

Text A 参考译文

数据库基本概念

1. 数据库

数据库是信息的集合，这些信息被组织起来以便可以容易地被访问、管理和更新。数据库可以按照其内容分为以下几类：书籍目录数据库、全文本数据库、数字数据库和图像数据库。

在计算领域中，数据库有时也按照其组织方法来分类。当前最流行的方法就是关系数据库，即一个定义数据以便数据可以用多种不同的方法来重新组织和访问的表格式数据库。分布式数据库是一个在网络中的许多不同的地方分布或复制的数据库。面向对象编程数据库是一个适合用对象类和子类定义数据的数据库。

计算机数据库通常包含数据记录或文件的集合，如销售业务、产品目录和库存以及客户概况。通常，数据库管理程序给用户提供控制读/写访问、产生报表和分析使用情况的能力。数据库和数据库管理程序在大型机系统中非常普遍，但也出现在更小的分布式工作站和中等规模的系统中，如出现在AS/400或个人计算机中。

2. 关系数据库

关系数据库是数据项的集合，这些数据项组织为正式描述的表格的一个集合，其中的数据可以用多种方式访问或调整而无须重新组织数据库表。关系数据库由E. F. Codd于1970年在IBM创造。

关系数据库的标准用户和应用程序接口是结构化查询语言(SQL). SQL语句既可用于对关系数据库进行交互式信息查询，也可用于收集报表信息。

除了相对容易建立和访问之外，关系数据库的主要优点是容易扩展。建立了原始数据库后，可以增加新的数据库类别而无须对现有所有应用进行修改。

关系数据库是包含预设种类中数据的表格的集合。每个表（有时也称为关系）按列包含一个或多个数据类。每行包括由列定义的类型的唯一数据项。例如，一个典型的商务订单输入数据库可以包括一个描述客户的表，该表列有客户姓名、地址、电话号码等。另一个表描述订单：产品、客户、日期、销售价格等。该数据库的用户可以获得所需要的数据库概况。例如，一个分部经理也许需要在某个日期之后购买产品的全部客户的概况或报表。同一公司的金融服务经理可以从相同的表中获得需要支付的账号报表。

建立一个关系数据库后，可以在一个数据列中定义可能值的域以及未来可以应用到这些值的约束。例如，一个潜在客户域最多可以允许有10个客户的名称，但限制在一个表中只能列出三个这样的客户。

关系数据库的定义会产生一个元数据表或对该表、列、域和约束的正式描述。

3. SQL

SQL(结构化查询语言)是一个标准语言，用来进行对数据库的交互式查询并更新数据库，如IBM的DB2、微软Access以及来自Oracle、Sybase和Computer Associates的数库产品。尽管SQL既是一个ANSI标准，也是一个ISO标准，但许多数据库产品支持对标准语言有专门扩展的SQL。请求的形式是命令行语言，可以进行选择、插入、更新、找出数据的位置等。它也有一个编程接口。

4. 数据库管理系统

数据库管理系统（DBMS）有时也叫作数据库管理器，是让一个或多个计算机用户建立和访问数据库中数据的程序。DBMS 管理用户查询（及来自其他程序的查询），这样用户和其他程序就不需要知道这些数据在存储介质中的物理位置，并且在多用户系统中，也不必知道还有谁可能正在访问这些数据。在处理用户查询时，DBMS 确保数据的完整性（也就是确保可以持续地被访问并且一直按照预先要求组织好）和安全性（确保只有那些有访问权的用户才可以访问这些数据）。最典型的 DBMS 是关系数据库管理系统（RDBMS）。一个标准的用户和程序接口是结构化查询语言（SQL）。一个更新的 DBMS 是面向对象数据库管理系统（ODBMS）。

DBMS 可以被看作一个文件管理器，它管理数据库中的数据而不是文件系统中的文件。在 IBM 的大型机操作系统中，非关系数据管理器曾经（并且现在也是，因为这些老的应用系统仍然在使用）以访问方法而知名。

DBMS 通常是数据库产品的固有部分。在 PC 上，微软的 Access 是单一或小组用户 DBMS 的一个流行范例。微软的 SQL Server 是适用于多用户（客户）数据库查询的一个范例。其他流行的 DBMS（顺便说一下，这些全部都是 RDBMS）是 IBM 的 DB2、Oracle 的数据库管理产品线以及 Sybase 的产品。

IBM 的信息管理系统（IMS）是最初的 DBMS 之一。DBMS 也可被像 IBM 的客户信息管理系统（CICS）这样的业务管理程序使用，或与其结合使用。

5. 分布式数据库

分布式数据库是数据库中的某些部分存储在网络中的多个计算机中的数据库。用户可以在他们自己的位置访问该数据库的一部分，这样他们可以访问与其工作相关的数据而不会影响其他人的工作。

6. DDBMS

DDBMS（分布式数据库管理系统）是一个集中应用程序，它管理一个分布式数据库，就像该数据库存储在同一计算机上一样。DDBMS 定期地同步所有数据，并且在多个用户必须访问相同数据的情况下，确保在一个地方对数据的更新和删除在其他地方存储的数据中会自动反映出来。

7. 字段

在数据库表中，字段是用于单一数据块的数据结构。字段组成为记录，包括表中与特定实体相关的全部信息。例如，在一个叫作客户联系信息的表中，电话号码可能是一行中的一个字段，该行也包含了其他字段，如街道地址和城市。记录构成了表行而字段构成了列。

8. 记录

在数据库中，记录（有时也叫作行）是表中与一个特定实体相关的一组字段。例如，在一个叫作客户联系信息的表中，一行可能包括这样的字段：标识号、名字、街道地址、城市、电话号码等。

9. 表

在关系数据库中，表（有时叫做文件）把单一主题的信息组成为行和列。例如，一个商用数据库通常包括一个客户信息表，该表把客户账号、地址、电话号码等存储为一系列的列。每个单一的数据块（如账号）是表中的字段。一列由单一字段的全部实体组成，如全部客户的电话号码。字段依次地被组织为记录，这就组成了信息的完整集合（如某一特定客户的信息集合），每个记录构成一行。这个规范处理过程决定了怎样将数据最有效地组织为表。

Text B

Data Warehousing

扫码听课文

1. What Is Data Warehousing

A data warehousing (DW) is process for collecting and managing data from various sources to provide meaningful business insights. A data warehouse is typically used to connect and analyze business data from heterogeneous sources. The data warehouse is the core of the BI system which is built for data analysis and reporting.

It is a blend of technologies and components which aids the strategic use of data. It is electronic storage of a large amount of information by a business which is designed for query and analysis instead of transaction processing. It is a process of transforming data into information and making it available to users in a timely manner to make a difference.

2. How Does Data Warehouse Work

A data warehouse works as a central repository where information arrives from one or more data sources. Data flows into a data warehouse from the transactional system and other relational databases.

Data may be structured, semi-structured and unstructured data.

The data is processed, transformed, and ingested so that users can access the processed data in the data warehouse through business intelligence tools, SQL clients, and spreadsheets. A data warehouse merges information coming from different sources into one comprehensive database.

By merging all of this information in one place, an organization can analyze its customers more holistically. This helps to ensure that it has considered all the information available. Data warehousing makes data mining possible. Data mining is looking for patterns in the data that may lead to higher sales and profits.

3. The Types of Data Warehouse

3.1 Enterprise Data Warehouse

Enterprise data warehouse is a centralized warehouse. It provides decision support service across the enterprise. It offers a unified approach for organizing and representing data. It also provide the ability to classify data according to the subject and give access according to those divisions.

3.2 Operational Data Store

Operational data store, which is also called ODS, are nothing but data store required when neither data warehouse nor OLTP systems support organizations reporting needs. In ODS, Data warehouse is refreshed in real time. Hence, it is widely preferred for routine activities like storing records of the employees.

3.3 Data Mart

A data mart is a subset of the data warehouse. It is specially designed for a particular line of business, such as sales and finance. In an independent data mart, data can collect directly from sources.

4. The General Stages of Data Warehouse

Organizations originally started relatively simple use of data warehousing. However, over time, more sophisticated use of data warehousing begun. The following are general stages of use of the data warehouse.

4.1 Offline Operational Database

In this stage, data is just copied from an operational system to another server. In this way, loading, processing, and reporting of the copied data do not impact the operational system's performance.

4.2 Offline Data Warehouse

Data in the data warehouse is regularly updated from the operational database. The data in data warehouse is mapped and transformed to meet the data warehouse objectives.

4.3 Real Time Data Warehouse

In this stage, data warehouses are updated whenever any transaction takes place in operational database. For example, airline or railway booking system.

4.4 Integrated Data Warehouse

In this stage, data warehouses are updated continuously when the operational system performs a transaction. The data warehouse then generates transactions which are passed back to the operational system.

5. The Components of Data Warehouse

5.1 Load Manager

Load manager is also called the front component. It performs all the operations associated with the extraction and load of data into the warehouse. These operations include transformations to prepare the data for entering into the data warehouse.

5.2 Warehouse Manager

Warehouse manager performs operations associated with the management of the data in the warehouse. It performs operations like analysis of data to ensure consistency, creation of indexes and views, generation of denormalization and aggregations, transformation and merging of source data and archiving and backup data.

5.3 Query Manager

Query manager is also known as backend component. It performs all the operations related to the management of user queries. The operations of this data warehouse components are direct queries to the appropriate tables for scheduling the execution of queries.

6. Who Needs Data Warehouse

Data warehouse is needed for all types of users like:

- Decision makers who rely on mass amount of data.
- Users who use customized, complex processes to obtain information from multiple data sources.
- It is also used by the people who want simple technology to access the data.
- It's also essential for those people who want a systematic approach for making decisions.
- If the user wants fast performance on a huge amount of data which is a necessity for reports, grids or charts, then data warehouse proves useful.
- Data warehouse is a first step if you want to discover "hidden patterns" of data-flows and groupings.

7. The Steps to Implement Data Warehouse

The best way to address the business risk associated with a data warehouse implementation is to employ a three-prong strategy as below:

- Enterprise strategy: Here we identify technical including current architecture and tools. We also identify facts, dimensions, and attributes. Data mapping and transformation is also passed.
- Phased delivery: Data warehouse implementation should be phased based on subject areas. Related business entities like booking and billing should be first implemented and then integrated with each other.
- Iterative prototyping: Rather than a big bang approach to implementation, the data warehouse should be developed and tested iteratively.

Best practices to implement a data warehouse:

- Make a plan to test the consistency, accuracy, and integrity of the data.
- The data warehouse must be well-integrated, well-defined and time-stamped.
- While designing data warehouse make sure you use the right tool, stick to life cycle, take care about data conflicts and ready to learn from your mistakes.
- Never replace operational systems and reports.
- Don't spend too much time on extracting, cleaning and loading data.
- Ensure to involve all stakeholders including business personnel in data warehouse implementation process. Establish that data warehousing is a joint/ team project. You don't want to create data warehouse that is not useful to the end users.
- Prepare a training plan for the end users.

8. The Advantages of Data Warehouse

- It allows business users to quickly access critical data from some sources all in one place.
- It provides consistent information on various cross-functional activities. It is also supporting ad-hoc reporting and query.
- It helps to integrate many sources of data to reduce stress on the production system.
- It helps to reduce total turnaround time for analysis and reporting.
- Restructuring and integration make it easier for the user to use for reporting and analysis.
- It allows users to access critical data from the number of sources in a single place. Therefore, it saves user's time of retrieving data from multiple sources.
- It stores a large amount of historical data. This helps users to analyze different time periods and trends to make future predictions.

9. The Disadvantages of Data Warehouse

- Data warehouse is not an ideal option for unstructured data.
- Creation and implementation of data warehouse is surely a time consuming affair.
- Data warehouse can be outdated relatively quickly.

- It is difficult to make changes in data types and ranges, data source schema, indexes, and queries.
- The data warehouse may seem easy, but actually, it is too complex for the average users.
- Despite best efforts at project management, data warehousing project scope will always increase.
- Sometime warehouse users will develop different business rules.
- Organizations need to spend lots of their resources for training and implementation purpose.

10. The Future of Data Warehousing

Data warehouses are here for the long term. People have invested a lot in building them and many people and business functions depend on them. But the sustainability demands that we rethink the data warehouse. Data warehouse architecture can no longer stand alone. We must think of the purpose, placement, and positioning of the data warehouse in a broader data management architecture.

Data warehouse faces many challenges. It doesn't scale well, it has performance bottlenecks, it can be difficult to change, and it doesn't work well for big data. In the future we'll need to consider cloud data warehousing, data warehousing with Hadoop, data warehouse automation as well as architectural modernization.

New Words

warehousing	[ˈweəhaʊzɪŋ]	*n.*仓库
insight	[ˈɪnsɑɪt]	*n.*洞察力，洞见
heterogeneous	[ˌhetərəˈdʒiːnɪəs]	*adj.*各种各样的
repository	[rɪˈpɒzətərɪ]	*n.*仓库，储藏室
transactional	[træn'zækʃənəl]	*adj.*交易的，业务的
semi-structured	[ˈsemə'strʌktʃəd]	*adj.*半结构的
unstructured	[ʌnˈstrʌktʃəd]	*adj.*无结构的，非结构的
ingest	[ɪnˈdʒest]	*vt.*获取；吸收
merge	[mɜːdʒ]	*v.*（使）混合；相融；融入
holistically	[həʊˈlɪstɪklɪ]	*adv.*整体地；全面地
enterprise	[ˈentəpraɪz]	*n.*企（事）业单位
subject	[ˈsʌbdʒɪkt]	*n.*主题，话题
refresh	[rɪˈfreʃ]	*v.*刷新
offline	[ˌɒfˈlaɪn]	*adj.*未连线的；未联机的；脱机的；离线的
regularly	[ˈregjʊləlɪ]	*adv.*有规律地，定期地

consistency	[kənˈsɪstənsɪ]	*n.*前后一致，连贯；符合
normalization	[ˌnɔ:məlɑɪ'zeɪʃn]	*n.*正常化；标准化；正态化
systematic	[ˌsɪstəˈmætɪk]	*adj.*系统的，有规则的
chart	[tʃɑ:t]	*n.*图表
data-flow	[ˈdeɪtə-fləʊ]	*n.*数据流
iteratively	['ɪtərətɪvlɪ]	*adv.*重复地，反复地，迭代地
accuracy	[ˈækjʊrəsɪ]	*n.*精确（性），准确（性）
extracting	[ɪks'træktɪŋ]	*n.*提取，提炼
establish	[ɪˈstæblɪʃ]	*vt.*建立，创建；确立
ad-hoc	['æd'hɒk]	*n.*为某种目的设置的，特别的
sustainability	[səˌstenə'bɪlɪtɪ]	*n.*持续性
demand	[dɪˈmɑ:nd]	*v.&n.*需求，要求
rethink	[ˌri:ˈθɪŋk]	*v.&n.*重新考虑，反思
placement	[ˈpleɪsmənt]	*n.*安置，放置；定位
bottleneck	[ˈbɒtlnek]	*n.*瓶颈，阻碍
modernization	[ˌmɒdənɑɪ'zeɪʃn]	*n.*现代化，现代化的事物

Phrases

a blend of...	……的混合
transaction processing	业务处理
transform ... into ...	把……转变成……，把……转换成……
data mining	数据挖掘
look for	寻找
data mart	数据集市
real time	实时
associate with	与……联系
hidden pattern	隐藏的模式
stick to	遵守；坚持
turnaround time	提交时间，周转时间，回复时间
project management	项目管理
cloud data warehousing	云数据仓库

Abbreviations

DW (Data Warehousing)	数据仓库
BI (Business Intelligence)	商业智能，商务智能

ODS (Operational Data Store)　　操作型数据存储
OLTP (On-Line Transaction Processing)　　联机事务处理

Text B 参考译文

数 据 仓 库

1. 什么是数据仓库

数据仓库（DW）是用于收集和管理来自各种来源的数据以提供有意义的业务见解的过程。数据仓库通常用于连接和分析来自异构源的业务数据。数据仓库是商务智能系统的核心，它是为数据分析和报告而构建的。

它是技术和部件的组合，有助于战略性地使用数据。它是企业为查询和分析而不是事务处理而设计，它电子化地存储了大量信息。这是将数据转换为信息并及时将其提供给用户以产生影响的过程。

2. 数据仓库如何工作

数据仓库用作中央存储库，信息来自一个或多个数据源。事务系统和其他关系数据库的数据流入数据仓库。

数据可以是结构化、半结构化和非结构化数据。

数据经过处理、转换和提取，用户就可以通过商业智能工具、SQL 客户端和电子表格访问数据仓库中已处理的数据。数据仓库将来自不同来源的信息合并到一个综合数据库中。

通过将所有这些信息合并到一个位置，组织可以更全面地分析其客户。这有助于确保考虑所有的可用信息。数据仓库使数据挖掘成为可能。数据挖掘是寻找可能带来更高的销售量和利润额的数据模式。

3. 数据仓库的类型

3.1　企业数据仓库

企业数据仓库是一个集中式仓库。它为整个企业提供决策支持服务。它提供了用于组织和表示数据的统一方法。它还提供了根据主题对数据进行分类并根据这些划分进行访问的功能。

3.2　操作型数据存储

当数据仓库和 OLTP 系统都不支持组织报告需求时，操作型数据存储（也称为 ODS）就只存储需要的数据。在 ODS 中，数据仓库是实时刷新的。因此，它被广泛用于日常活动，例如存储员工记录。

3.3 数据集市

数据集市是数据仓库的子集。它是为特定业务行业（例如销售和财务）专门设计的。在独立的数据集市中，数据可以直接从源收集。

4. 数据仓库的一般阶段

组织最初开始相对简单地使用数据仓库。但是，随着时间的流逝，开始更加复杂地使用数据仓库。以下是使用数据仓库的一般阶段。

4.1 离线运营数据库

在此阶段，数据只是从运营系统复制到另一台服务器。这样，加载、处理和报告复制的数据不会影响操作系统的性能。

4.2 离线数据仓库

数据仓库中的数据会定期从运营数据库中更新。映射并转换数据仓库中的数据，以满足数据仓库的目标。

4.3 实时数据仓库

在此阶段，每当运营数据库中发生任何事务时，都会更新数据仓库。例如，航空公司或铁路订票系统。

4.4 集成数据仓库

在此阶段，当运营系统执行事务时，数据仓库将不断更新。然后，数据仓库将生成事务，这些事务将被传回运营系统。

5. 数据仓库的组成

5.1 负载管理器

负载管理器也称为前端组件。它执行与提取数据并将数据加载到仓库相关的所有操作。这些操作包括为准备要输入到数据仓库中的数据而进行的转换。

5.2 仓库管理器

仓管理器执行与仓库中数据管理相关的操作。它执行的操作有：分析数据以确保一致性，创建索引和视图，生成非规范化和聚合，转换和合并源数据以及归档和备份数据等。

5.3 查询管理器

查询管理器也称为后端组件。它执行与用户查询管理相关的所有操作。该数据仓库组件的操作是对适当表的直接查询，以安排执行查询。

6. 谁需要数据仓库

所有类型的用户都需要数据仓库，例如：

- 依赖大量数据的决策者。
- 使用定制的复杂过程从多个数据源获取信息的用户。
- 希望使用简单技术访问数据的人也可以使用它。
- 对于那些希望采用系统方法进行决策的人们来说，这也是至关重要的。
- 如果用户希望在海量数据上实现快速性能，而这些数据对于报表、表格或图表是必不可少的，则证明数据仓库非常有用。
- 如果要发现数据流和分组的“隐藏模式”，则数据仓库是第一步。

7. 实施数据仓库的步骤

解决与数据仓库实施相关的业务风险的最佳方法是采用以下三种策略：

- 企业战略：我们确定技术，包括当前的体系结构和工具。我们还确定事实、维度和属性。也要进行数据映射和转换。
- 分阶段交付：应根据主题领域分阶段实施数据仓库。相关业务实体（例如预订和计费）应首先实现，然后再相互集成。
- 迭代原型：应该对数据仓库进行迭代开发和测试，而不是采用大规模的实现方法。

实施数据仓库的最佳实践：

- 制定计划以测试数据的一致性、准确性和完整性。
- 数据仓库必须很好地集成、定义良好并带有时间戳。
- 在设计数据仓库时，请确保使用正确的工具，遵循生命周期，注意防止数据冲突并准备从错误中学习。
- 切勿更换运营和报告系统。
- 不要在提取、清理和加载数据上花费太多时间。
- 确保让所有利益相关者（包括业务人员）参与数据仓库实施过程。确定数据仓库是一个联合/团队项目。不要创建对最终用户无用的数据仓库。
- 为最终用户准备培训计划。

8. 数据仓库的优势

- 它使业务用户可以只在一处就能快速地访问来自某些数据源的关键数据。
- 它提供有关各种跨职能活动的一致信息。它还支持特别报告和查询。
- 它有助于集成许多数据源，以减轻生产系统的压力。

- 它有助于减少分析和报告的总提交时间。
- 重组和集成使用户可以更轻松地用它来报告和分析。
- 它使用户可以在一处访问来自多个数据源的关键数据。因此，它节省了用户从多个来源检索数据的时间。
- 它存储大量历史数据。这可以帮助用户分析不同的时间段和趋势以预测未来。

9. 数据仓库的缺点

- 数据仓库不是非结构化数据的理想选择。
- 创建和实施数据仓库无疑是一项耗时的工作。
- 数据仓库相对来说过时较快。
- 很难更改数据类型和范围、数据源模式、索引和查询。
- 数据仓库看似容易，但实际上对于普通用户而言它太复杂了。
- 尽管尽最大努力进行项目管理，但数据仓库项目范围将不断扩大。
- 有时仓库用户将制定不同的业务规则。
- 组织需要将大量资源用于培训和实施。

10. 数据仓库的未来

数据仓库会长期存在。人们在构建数据仓库上投入了大量资金，许多人和业务职能都依赖于它们。但是可持续性要求我们重新考虑数据仓库。数据仓库架构不再单独存在。我们必须考虑更广泛的数据管理体系结构中数据仓库的目的、安排和位置。

数据仓库面临许多挑战。它不能很好地扩展，它具有性能瓶颈，很难更改，并且不适用于大数据。将来，我们将需要考虑云数据仓库、使用 Hadoop 的数据仓库、数据仓库自动化以及架构现代化。

Exercises

[Ex. 1] Decide whether the following statement are True or False according to Text A.

1. A database is a collection of organized information.

2. A relational database is a tabular database in which data is defined so that it can be reorganized and accessed in a number of different ways.

3. A distributed database is one that can be dispersed or replicated at certain points in a network.

4. An object-oriented programming database is one that is congruent with the data defined in object classes and subclasses.

5. Databases and database managers are prevalent only in large mainframe systems.

6. The relational database was invented by E. F. Codd at IBM in 1970.

7. A relational database has the important advantage of being easy to extend except being relatively difficult to create and access.

8. SQL (structured query language) is a standard language for making interactive queries from a database and updating a database.

9. A database management system (DBMS), sometimes just called a database manager, is a program that lets only one computer users create and access data in a database.

10. The most typical DBMS is a distributed database management system.

11. A newer kind of DBMS is the object-oriented database management system (ODBMS).

12. A DBMS can be thought of as a file manager that manages data in databases.

13. A distributed database is a database in which portions of the database are stored on one computer.

14. A DDBMS is a centralized application that manages a relational database on the same computer.

15. In a database table, a field is a data structure for a single piece of data.

16. In a database, a record (sometimes called a row) is a group of files within a table that are relevant to a specific entity.

17. In a relational database, a table (sometimes called a file) organizes the information about a single topic into rows and columns.

18. The records make up the columns and the fields make up the table rows.

[Ex. 2] Answer the following questions according to Text B.

1. What is a data warehousing?

2. What does a data warehouse work as?

3. What are the types of data warehouse mentioned in the passage?

4. What are the general stages of use of the data warehouse?

5. What are the components of data warehouse?

6. What is a first step if you want to discover "hidden patterns" of data-flows and groupings?

7. What is the best way to address the business risk associated with a data warehouse implementation?

8. What is the last but one advantage of data warehouse listed in the passage?

9. What is the fourth disadvantage of data warehouse listed in the passage?

10. What will we need to consider in the future?

[Ex. 3] Translate the following terms or phrases from English into Chinese and vice versa.

1. distributed database	1. ________
2. cloud data warehousing	2. ________

3. data mining ________ 3. ________
4. data mart ________ 4. ________
5. real time ________ 5. ________
6. *n.*方法，步骤，途径，通路 ________ 6. ________
7. *adj.*完备的，完全的，完成的 ________ 7. ________
8. *n.*域；字段 ________ 8. ________
9. *v.*确保 ________ 9. ________
10. *v.*同步 ________ 10. ________

[Ex. 4] Translate the following sentences into Chinese.

DDL (Data Definition Language)

Data Definition Language is used to define the database structure or schema. DDL is also used to specify additional properties of the data. The storage structure and access methods are used by the database system by a set of statements in a special type of DDL called a data storage and definition language. These statements define the implementation details of the database schema, which are usually hidden from the users. The data values stored in the database must satisfy certain consistency constraints.

For example, suppose the university requires that the account balance of a department must never be negative. The DDL provides facilities to specify such constraints. The database system checks these constraints every time the database is updated. In general, a constraint can be an arbitrary predicate pertaining to the database. However, arbitrary predicates may be costly to the test. Thus, the database system implements integrity constraints that can be tested with minimal overhead.

1. Domain Constraints

A domain of possible values must be associated with every attribute (for example, integer types, character types, date/time types). Declaring an attribute to be of a particular domain acts as the constraints on the values that it can take.

2. Assertions

An assertion is any condition that the database must always satisfy. Domain constraints and integrity constraints are special form of assertions.

3. Authorization

We may want to differentiate among the users as far as the type of access they are permitted on various data values in database. These differentiation are expressed in terms of Authorization. The most common are :

- Read authorization—which allows reading but not modification of data;

- Insert authorization—which allow insertion of new data but not modification of existing data;
- Update authorization—which allows modification, but not deletion.

[Ex. 5] Fill in the blanks with the words given below.

transactional	access	website	professionals	previous
affected	calculation	budgeting	delete	mining

OLTP and OLAP

1. Online Transaction Processing (OLTP)

OLTP databases are meant to be used to do many small transactions, and usually serve as a "single source of storage". An example of OLTP system is online movie ticket booking ____1____. Suppose two persons at the same time want to book the same seat for the same movie for the same movie timing then in this case whoever will complete the transaction first will get the ticket. The key thing to note here is that OLTP systems are designed for ____2____ priority instead data analysis.

Benefits of using OLTP services:

- The main benefit of using OLTP services is it responds to its user actions immediately as it can process query very quickly.
- OLTP services allows its user to perform operations like read, write and ____3____ data quickly.

Drawbacks of OLTP services:

- The major problem with the OLAP services is it is not fail-safe. If there is hardware failure, then online transactions get ____4____.
- OLTP allow users to ____5____ and change the data at the same time which cause unprecedented situation.

2. Online Analytic Processing (OLAP)

OLAP databases on the other hand are more suited for analytics, data ____6____, less queries but they are usually bigger (they operate on more data). We can say that any data warehouse system is an OLAP system. Many company compare their sales of current month with ____7____ month to keep trace of business. Here company compare the sales and keep the result in another location, which is a separate database. Here company uses OLAP databases.

Benefits of using OLAP services:

- The main benefit of using OLAP services is it helps to keep trace of consistency and ____8____.
- OLAP builds one single platform where we can store planning, analysis and ____9____ for business analytics.

- With the OLAP as service, we can easily apply security restrictions to protect data.

Drawbacks of OLAP services:

- The major problem with the OLAP services is it always needs IT ___10___ to handle the data because OLAP tools require a complicated modeling procedure.
- OLAP tools need cooperation between people of various departments, which leads dependency problem.

Online Resources

二维码	内　容
	计算机专业常用语法（5）：过去分词
	在线阅读（1）：The Difference Between a Data Warehouse and a Database（数据仓库与数据库的不同）
	在线阅读（2）：Cloud Database Deployment Models（云数据库开发模型）

Unit 6

Object Oriented Design

Text A

Object-Oriented Technology

扫码听课文

1. Object-Oriented Analysis

Object–oriented analysis (OOA) is the procedure of identifying software engineering requirements and developing software specifications in terms of a software system's object model, which comprises of interacting objects.

The main difference between object-oriented analysis and other forms of analysis is that in object-oriented approach, requirements are organized around objects, which integrate both data and functions. They are modelled after real-world objects that the system interacts with. In traditional analysis methodologies, the two aspects—functions and data—are considered separately.

The primary tasks in object-oriented analysis are:

- Identifying objects.
- Organizing the objects by creating object model diagram.
- Defining the internals of the objects, or object attributes.
- Defining the behavior of the objects, i.e., object actions.
- Describing how the objects interact.

The common models used in OOA are use cases and object models.

2. Object-Oriented Design

Object–oriented design (OOD) involves implementation of the conceptual model produced during object-oriented analysis. In OOD, concepts in the analysis model, which are technology–independent, are mapped onto implementing classes, constraints are identified and interfaces are designed, resulting in a model for the solution domain.

The implementation details generally include:

- Restructuring the class data (if necessary).
- Implementation of methods, i.e., internal data structures and algorithms.

- Implementation of control.
- Implementation of associations.

3. Object-Oriented Programming

Object-oriented programming (OOP) is a programming paradigm based upon objects (having both data and methods) that aims to incorporate the advantages of modularity and reusability. Objects, which are usually instances of classes, are used to interact with one another to design applications and computer programs.

The important features of object–oriented programming are:

- Bottom–up approach in program design.
- Programs organized around objects, grouped in classes.
- Focus on data with methods to operate upon object's data.
- Interaction between objects through functions.
- Reusability of design through creation of new classes by adding features to existing classes.

Some examples of object-oriented programming languages are C++, Java, Smalltalk, Delphi, C#, Perl, Python, Ruby, and PHP.

4. Objects and Classes

4.1 Object

An object is a real-world element in an object–oriented environment that may have a physical or a conceptual existence. Each object has:

- Identity that distinguishes it from other objects in the system.
- State that determines the characteristic properties of an object as well as the values of the properties that the object holds.
- Behavior that represents externally visible activities performed by an object in terms of changes in its state.

Objects can be modelled according to the needs of the application. An object may have a physical existence, like a customer, a car, etc.; or an intangible conceptual existence, like a project, a process, etc.

4.2 Class

A class represents a collection of objects having same characteristic properties that exhibit common behavior. It gives the blueprint or description of the objects that can be created from. Creation of an object as a member of a class is called instantiation. Thus, object is an instance of a class.

The constituents of a class are:

- A set of attributes for the objects that are to be instantiated from the class. Generally, different objects of a class have some differences in the values of the attributes. Attributes are often referred to as class data.

- A set of operations that portray the behavior of the objects of the class. Operations are also referred to as functions or methods.

Let us consider a simple class, Circle, that represents the geometrical figure circle in a two–dimensional space. The attributes of this class can be identified as follows:

- *x*–coord, to denote *x*–coordinate of the center
- *y*–coord, to denote *y*–coordinate of the center
- *a*, to denote the radius of the circle

Some of its operations can be defined as follows:

- findArea( ), method to calculate area
- findCircumference( ), method to calculate circumference
- scale( ), method to increase or decrease the radius

During instantiation, values are assigned for some of the attributes. If we create an object my_circle, we can assign values like *x*-coord : 2, *y*-coord : 3, and *a* : 4 to depict its state. Now, if the operation scale( ) is performed on my_circle with a scaling factor of 2, the value of the variable a will become 8. This operation brings a change in the state of my_circle, i.e., the object has exhibited certain behavior.

5. Encapsulation and Data Hiding

5.1 Encapsulation

Encapsulation is the process of binding both attributes and methods together within a class. Through encapsulation, the internal details of a class can be hidden from outside. It permits the elements of the class to be accessed from outside only through the interface provided by the class.

5.2 Data Hiding

Typically, a class is designed such that its data (attributes) can be accessed only by its class methods and insulated from direct outside access. This process of insulating an object's data is called data hiding or information hiding.

In the class Circle, data hiding can be incorporated by making attributes invisible from outside the class and adding two more methods to the class for accessing class data, namely:

- setValues( ), method to assign values to *x*-coord, *y*-coord, and *a*
- getValues( ), method to retrieve values of *x*-coord, *y*-coord, and *a*

Here the private data of the object my_circle cannot be accessed directly by any method that is not encapsulated within the class Circle. It should instead be accessed through the methods setValues( ) and getValues( ).

6. Message Passing

An application requires a number of objects interacting in a proper manner. Objects in a system may communicate with each other using message passing. Suppose a system has two objects: obj1 and obj2. The object obj1 sends a message to object obj2, if obj1 wants obj2 to execute one of its methods.

The features of message passing are:

- Message passing between two objects is generally unidirectional.
- Message passing enables all interactions between objects.
- Message passing essentially involves invoking class methods.
- Objects in different processes can be involved in message passing.

7. Inheritance

Inheritance is the mechanism that permits new classes to be created out of existing classes by extending and refining its capabilities. The existing classes are called the base classes/parent classes/superclasses, and the new classes are called the derived classes/child classes/subclasses. The subclass can inherit or derive the attributes and methods of the superclass(es) provided that the superclass allows so. Besides, the subclass may add its own attributes and methods and may modify any of the superclass methods. Inheritance defines an "is – a" relationship.

From a class Mammal, a number of classes can be derived from, such as human, cat, dog, cow, etc. Humans, cats, dogs, and cows all have the distinct characteristics of mammals. In addition, each has its own particular characteristics. It can be said that a cow "is – a" mammal.

Types of inheritance:

- Single inheritance —A subclass derives from a single superclass.
- Multiple inheritance —A subclass derives from more than one superclasses.
- Multilevel inheritance —A subclass derives from a superclass which, in turn, is derived from another class and so on.
- Hierarchical inheritance —A class has a number of subclasses, each of which may have subsequent subclasses, continuing for a number of levels, so as to form a tree structure.
- Hybrid inheritance —A combination of multiple and multilevel inheritance so as to form a lattice structure.

8. Polymorphism

Polymorphism is originally a Greek word that means the ability to take multiple forms. In object-oriented paradigm, polymorphism implies using operations in different ways, depending upon the instance they are operating upon. Polymorphism allows objects with different internal structures to have a common external interface. Polymorphism is particularly effective while implementing inheritance.

Let us consider two classes, Circle and Square, each with a method findArea( ). Though the name and purpose of the methods in the classes are same, the internal implementation, i.e., the procedure of calculating area is different for each class. When an object of class Circle invokes its findArea( ) method, the operation finds the area of the circle without any conflict with the findArea( ) method of the Square class.

9. Generalization and Specialization

Generalization and specialization represent a hierarchy of relationships between classes, where subclasses inherit from superclasses.

9.1 Generalization

In the generalization process, the common characteristics of classes are combined to form a class in a higher level of hierarchy, i.e., subclasses are combined to form a generalized super-class. It represents an "is – a kind of" relationship. For example, "car is a kind of land vehicle", or "ship is a kind of water vehicle".

9.2 Specialization

Specialization is the reverse process of generalization. Here, the distinguishing features of groups of objects are used to form specialized classes from existing classes. It can be said that the subclasses are the specialized versions of the superclass.

New Words

requirement	[rɪˈkwaɪəmənt]	*n.*需求，要求
object	[ˈɒbdʒɪkt]	*n.*对象；物体；目标；客体
methodology	[ˌmeθəˈdɒlədʒɪ]	*n.*一套方法；方法学；方法论
separately	[ˈseprətlɪ]	*adv.*分别地，分离地；个别地；分开，单独
attribute	[əˈtrɪbju:t]	*n.*属性，性质，特征
behavior	[bɪ'heɪvjə]	*n.*行为；态度
conceptual	[kənˈseptʃuəl]	*adj.*概念的，观念的
class	[klɑ:s]	*n.*类
restructure	[ˌri:ˈstrʌktʃə]	*v.*重构，重建，重组
association	[əˌsəuʃɪˈeɪʃn]	*n.*联合；联系
paradigm	[ˈpærədaɪm]	*n.*范式，范例，样式
incorporate	[ɪnˈkɔ:pəreɪt]	*vt.*包含；使混合 *vi.*包含；吸收；合并；混合
instance	[ˈɪnstəns]	*n.*实例，例子
bottom–up	[ˈbɒtəm-ʌp]	*adj.*自底向上的
interaction	[ˌɪntər'ækʃn]	*n.*交互，互动，互相影响
reusability	[rɪju:zə'bɪlɪtɪ]	*n.*可重用性
distinguish	[dɪˈstɪŋgwɪʃ]	*v.*区分，辨别
characteristic	[ˌkærəktəˈrɪstɪk]	*adj.*特有的；独特的 *n.*特性，特征，特色
property	[ˈprɒpətɪ]	*n.*特性；属性

intangible	[ɪnˈtændʒəbl]	*adj.*无形的；触不到的
exhibit	[ɪgˈzɪbɪt]	*vt.*呈现
instantiation	[ɪnstænʃɪ'eɪʃn]	*n.*实例化
constituent	[kənˈstɪtjʊənt]	*n.*成分，构成部分 *adj.*构成的，组成的
portray	[pɔːˈtreɪ]	*v.*表现；描述
geometrical	[ˌdʒiːə'metrɪkl]	*adj.*几何的，几何学的；成几何级数增加的
coordinate	[kəʊ'ɔːdɪnɪt]	*n.*坐标
denote	[dɪˈnəʊt]	*vt.*代表；意思是
circumference	[səˈkʌmfərəns]	*n.*圆周
depict	[dɪˈpɪkt]	*vt.*描绘，描述
encapsulation	[ɪnˌkæpsjʊˈleɪʃn]	*n.*封装
hide	[haɪd]	*v.*隐藏，隐蔽
permit	[pəˈmɪt]	*vt.*许可，准许 *n.*许可证，执照
insulate	[ˈɪnsjʊleɪt]	*vt.*使隔离；使绝缘
send	[send]	*v.*发出信息
essentially	[ɪˈsenʃəlɪ]	*adv.*本质上，根本上
inheritance	[ɪnˈherɪtəns]	*n.*继承；遗传
extend	[ɪkˈstend]	*v.*延伸，扩展
refine	[rɪˈfaɪn]	*vt.*提炼；改善
superclass	[ˈsuːpəklɑːs]	*n.*超类
subclass	[sʌbklɑːs]	*n.*子类
mammal	[ˈmæml]	*n.*哺乳动物
hybrid	[ˈhaɪbrɪd]	*adj.*混合的
combination	[ˌkɒmbɪˈneɪʃn]	*n.*组合，结合；联合体
polymorphism	[ˌpɒlɪ'mɔːfɪzəm]	*n.*多态性
imply	[ɪmˈplaɪ]	*v.*说明，表明；暗示；意味；隐含
generalization	[ˌdʒenrəlaɪˈzeɪʃn]	*n.*泛化
specialization	[ˌspeʃəlaɪ'zeɪʃn]	*n.*特化
hierarchy	[ˈhaɪərɑːkɪ]	*n.*层次，分层

Phrases

object-oriented technology	面向对象技术
object-oriented approach	面向对象方法

object model diagram	对象模型图
use case	用例
object model	对象模型
analysis model	分析模型
be mapped onto	被映射到
a member of ...	……的一员
two–dimensional space	二维空间
be assigned for	分配给，赋值给
scaling factor	比例因子，缩放比例
data hiding	数据隐藏
information hiding	信息隐藏
message passing	消息传递
in a manner	用适当方式，在一定程度上
parent class	父类
single inheritance	单一继承
multiple inheritance	多重继承
multilevel inheritance	多级继承
derive from ...	由……起源；来自……
hierarchical inheritance	层次继承
hybrid inheritance	混合继承
lattice structure	网格结构

Abbreviations

OOA (Object–Oriented Analysis)	面向对象分析
OOD (Object–Oriented Design)	面向对象设计
OOP (Object-Oriented Programming)	面向对象编程

Text A 参考译文

面向对象技术

1. 面向对象分析

面向对象分析（OOA）是根据软件系统的对象模型（包括交互对象）识别软件工程需求

并制定软件规范的过程。

面向对象分析与其他形式的分析之间的主要区别在于，在面向对象的方法中，需求是围绕对象组织的，这些对象集成了数据和功能。它们根据与系统交互的实际对象建模。在传统的分析方法中，功能和数据这两个方面是分开考虑的。

面向对象分析的主要任务是：

- 识别对象。
- 通过创建对象模型图来组织对象。
- 定义对象或对象属性的内部。
- 定义对象的行为，即对象动作。
- 描述对象交互的方式。

OOA 中使用的常见模型是用例和对象模型。

2. 面向对象设计

面向对象设计（OOD）涉及实施在面向对象分析过程中产生的概念模型。在 OOD 中，分析模型中的概念与技术无关，它们被映射到实现类上，确定约束并设计接口，从而形成解决方案中的模型。

实施细节通常包括：

- 重组类数据（如有必要）。
- 方法的实现，即内部数据结构和算法。
- 实施控制。
- 实现关联。

3. 面向对象编程

面向对象编程（OOP）是一种基于对象（具有数据和方法）的编程范式，旨在融合模块化和可重用性的优点。对象通常是类的实例，用于彼此交互以设计应用程序和计算机程序。

面向对象编程的重要特征是：

- 程序设计中的自下而上的方法。
- 围绕对象组织的程序，按类分组。
- 以处理对象数据的方法聚焦数据。
- 以功能进行对象之间的交互。
- 向现有类添加功能来创建新类以实现设计的可重用性。

面向对象编程语言有以下示例：C ++、Java、Smalltalk、Delphi、C # 、Perl、Python、Ruby 和 PHP。

4. 对象和类

4.1 对象

对象是面向对象的环境中的现实世界元素，可能在物理上或概念上是存在的。每个对象都具有：

- 使其与系统中其他对象区分开的标识。
- 确定对象的特征属性以及该对象拥有的属性值的状态。
- 表示在状态变化时对象所进行的外部可见活动的行为。

可以根据应用程序的需求对对象建模。一个对象可能具有物理存在（例如客户、汽车等）或无形的概念性存在（例如项目、过程等）。

4.2 类

类表示具有相同特征属性的对象的集合，这些对象具有共同的行为。它给出了可以从中创建对象的蓝图或描述。创建作为类成员的对象称为实例化。因此，对象是类的实例。

类的组成部分包括：

- 将从类中实例化的对象的一组属性。通常，类的不同对象在属性值上有一些差异。属性通常称为类数据。
- 描述类对象行为的一组操作。操作也称为功能或方法。

让我们考虑一个简单的类 Circle，它代表二维空间中的几何图形圆。此类的属性可以标识如下：

- x 坐标，表示中心的 x 坐标。
- y 坐标，表示中心的 y 坐标。
- a，表示圆的半径。

其某些操作可以定义如下：

- findArea( )，用于计算面积的方法。
- findCircumference( )，用于计算周长的方法。
- scale( )，增加或减少半径的方法。

在实例化期间，为某些属性分配值。如果创建对象 my_circle，则可以分配值，如 *x*-coord：2，*y*-coord：3 和 a：4 来描述其状态。现在，如果在比例缩放因子为 2 的 my_circle 上执行了 scale( )操作，则变量 a 的值将变为 8。此操作将改变 my_circle 的状态，即对象已表现出一定的行为。

5. 封装和数据隐藏

5.1 封装

封装是将属性和方法绑定到类中的过程。通过封装，可以对外部隐藏类的内部细节。它允许仅通过类提供的接口从外部访问类的元素。

5.2 数据隐藏

通常，对一个类进行设计，以便只能通过其类方法访问其数据（属性），并且避免外部直接访问。这个隔离对象数据的过程称为数据隐藏或信息隐藏。

在类 Circle 中，可以通过以下方式融入数据隐藏：使属性从类外部不可见，并向该类添加另外两种方法以访问类数据，即：

- setValues( )，将值分配给 x 坐标、y 坐标和 a 的方法。
- getValues( )，用于获得 x 坐标、y 坐标和 a 值的方法。

在这里，对象 my_circle 的私有数据无法通过未封装在 Circle 类中的任何方法直接访问。相反，应通过 setValues( )和 getValues( )方法进行访问。

6. 消息传递

应用程序需要以适当方式进行交互的多个对象。系统中的对象可以使用消息传递相互通信。假设系统有两个对象：obj1 和 obj2。如果 obj1 希望 obj2 执行其某个方法，则对象 obj1 向对象 obj2 发送一条消息。

消息传递的特点包括：

- 在两个对象之间传递的消息通常是单向的。
- 消息传递使对象之间的所有交互成为可能。
- 消息传递主要涉及调用类方法。
- 消息传递中可能涉及不同进程中的对象。

7. 继承

继承是一种机制，可以通过扩展和完善其功能从现有类中创建新类。现有类称为基类/父类/超类，新类称为派生类/孩子类/子类。只要超类允许，子类可以继承或派生超类的属性和方法。此外，子类可以添加自己的属性和方法，并且可以修改任何超类方法。继承定义了“是- 一个”的关系。

从哺乳动物类中，可以衍生出许多类，例如人、猫、狗、牛等。人、猫、狗和牛都具有哺乳动物的独特特征。此外，每种动物都有其自己的特定特征。可以说，牛“是一个”哺乳动物。

继承类型：

- 单一继承：一个子类派生自单个超类。
- 多重继承：一个子类派生自一个以上的超类。
- 多级继承：一个子类从超类派生，而超类又从另一个类派生，依此类推。
- 层次继承：一个类具有多个子类，每个子类可能具有后续的子类，并连续多个级别，以形成树结构。

- 混合继承：多重继承和多级继承的组合，以形成网格结构。

8. 多态性

多态性最初是一个希腊词汇，表示能够具有多种形式。在面向对象的范例中，多态性意味着以不同的方式使用操作，具体取决于操作所基于的实例。多态性允许具有不同内部结构的对象具有共同的外部接口。多态在实现继承时特别有效。

让我们考虑两个类，即 Circle 和 Square，每个类都有一个 findArea()方法。尽管这些类中方法的名称和目的是相同的，但是内部实现，即每个类的面积计算过程是不同的。当类 Circle 的对象调用其 findArea( )方法时，该操作将查找圆的面积，而与类 Square 的 findArea( )方法没有任何冲突。

9. 泛化与特化

泛化和特化表示类之间关系的层次结构，其中子类继承自超类。

9.1 泛化

在泛化过程中，将类的共同特征组合起来，以形成更高层次结构中的类，即组合子类以形成广义的超类。它代表一种“是一种”关系。例如，“汽车是一种陆地车辆”，或“船是一种水上车辆”。

9.2 特化

特化是泛化的逆过程。在这里，用对象组的显著特征从现有类中形成特殊类。可以说子类是超类的特化版本。

Text B

扫码听课文

Computer Programmer

1. What Is a Computer Programmer

A computer programmer, or coder, is someone who writes computer software. The term computer programmer can refer to a specialist in one area of computer programming or to a generalist who writes code for many kinds of software.

The term programmer can be used to refer to a software developer, software engineer, computer scientist, or software analyst. However, members of these professions typically possess other software engineering skills beyond programming. For this reason, the term programmer is sometimes considered an oversimplification of these other professions.

2. What Does a Computer Programmer Do

The 21st century has brought in an extraordinary amount of technological progress. In the centre of this modern technology sits computer programmers, with the technological skills to create and navigate any new projects that may come their way. It's the job of computer programmers to take designs created by software developers and engineers and turn them into sets of instructions that computers can follow. These instructions result in the social media platforms, word processing programs, browsers, and more that people use every day.

There is an ongoing debate on the extent to which the writing of programs is an art, a craft, or an engineering discipline. In general, good programming is considered to be the measured application of all three, with the goal of producing an efficient software solution.

The discipline differs from many other technical professions in that a computer programmer, in general, does not need to be licensed or pass any standardized (or governmentally regulated) certification tests in order to call themselves a "programmer" or even a "software engineer".

A computer programmer figures out the process of designing, writing, testing, debugging/troubleshooting and maintaining the source code of computer programs. This source code is written in a programming language so the computer can "understand" it. The code may be a modification of an existing source or something completely new.

The purpose of programming is to create a program that produces a certain desired behaviour (customization). The process of writing source code often requires expertise in many different subjects, including knowledge of the application domain, specialized algorithms, and formal logic.

The computer programmer also designs a graphical user interface (GUI) so that non-technical users can use the software through easy, point-and-click menu options. The GUI acts as a translator between the user and the software code.

Some, especially those working on large projects that involve many computer programmers, use CASE (computer-aided software engineering) tools to automate much of the coding process. These tools enable a programmer to concentrate on writing the unique parts of a program. A programmer working on smaller projects will often use "programmer environments", or applications that increase productivity by combining compiling, code walk-through, code generation, test data generation, and debugging functions.

A computer programmer will also use libraries of basic code that can be modified or customized for a specific application. This approach yields more reliable and consistent programs and increases programmers' productivity by eliminating some routine steps. The computer programmer will also be responsible for maintaining the program's health.

As software design has continued to advance, and some programming functions have become automated, computer programmers have begun to assume some of the responsibilities that were once performed only by software engineers. As a result, some computer programmers now assist software engineers in identifying user needs and designing certain parts of computer programs.

3. Types of Computer Programmers

The products we use everyday, such as our computer, our smartphone, and our car, wouldn't be able to do what we ask them to do if it weren't for computer programmers. Computer programming is a very detail-oriented profession. Therefore, programmers are required to focus on code for long periods of time without losing focus or losing track of their progress. Often small but critical code issues can have a big impact technically, and can prevent a program from operating correctly. The ability to detect and rectify small discrepancies as quickly as possible is essential in order to solve issues.

3.1 Computer Hardware Programmer

Computers have their own machine language that they are able to understand and take instructions from. Computer hardware programmers write these instructions in a specific machine language (code) so that a computer knows what to do when someone presses the power button, types on the keyboard, or uses the mouse. They also write code so that text can be displayed when the computer turns on. Computer programs (a collection of instructions) are put in permanent memory storage so that as soon as someone turns on their computer, programmed information is displayed that prompts the user to choose what the computer does next.

Computer hardware programmers are also involved in researching, designing, developing, and testing computer equipment.

Various job tasks for computer hardware programmers: write detailed functional specifications for the hardware development process; build, test, and modify product prototypes using models; design, analyze and test performance of electrical, electronic, and computer equipment; evaluate interface between hardware and software; evaluate operational and performance requirements; prepare designs, determine specifications and determine operational plans; design and develop computer hardware (CPU's, support logic, microprocessors, circuits, printers and disk drives); monitor functioning and make necessary modifications; monitor processes for compliance with standards; recommend technical design or process changes to improve performance; store, retrieve, and manipulate data for analysis; Analyze user needs and recommend appropriate hardware.

Careers related to computer hardware programmer: automation engineer, computer architect, computer engineer, computer hardware designer, computer hardware developer, computer installation engineer, electronics engineer, field service engineer, hardware design engineer, hardware engineer, information technology consultant (IT consultant), network engineer, systems engineer, systems integration engineer, telecommunications engineer.

3.2 Web Developer

Computer programmers that design, create, and modify the millions of websites found on the Internet are called web developers. These types of computer programmers use software that allow them the ability to dictate what kinds of functions people are able to do when they access a website.

Web developers start by analyzing a user's needs before designing and structuring a website. They also add applicable graphics, audio, and video components if needed. Even a simple blog needs a web developer that can design the structure, function and the information that an audience can see.

Not only are web developers responsible for the look of a website, but they are also responsible for its performance, capacity, and sometimes content creation as well. Common programming languages for websites include HyperText Markup Language, JavaScript, Drupal, WordPress, and Joomla.

Careers related to web developer: Front end web developer, internet architect, php web developer, usability specialist, user experience designer, user interface developer, web applications developer, web architect, web page developer, web programmer, website developer, website specialist, web specialist, webmaster.

Various job tasks for web developers: analyze user needs, select programming languages, design tools, or applications, create web models or prototypes, write computer programming code, design, build, edit, or maintain websites, use content creation tools, management tools, and digital media, develop and document style guidelines for website content, perform or direct website updates, register websites with search engines, prioritize needs, resolve tech issues, and develop content criteria, back up files to local directories to prevent loss of information, resolve software problems and troubleshoot issues, ensure code is valid and meets industry standards, ensure code is compatible with browsers, devices, or operating systems, update knowledge of current web technologies and practices, develop test schedule to test performance, create and develop databases that support web applications and websites, develop and integrate e-commerce strategies and marketing strategies, provide technical support for computer network issues, design and implement firewalls or message encryption, develop website maps, application models, image templates, or page templates, prepare graphics or other visual representations of information, manage links to and from other websites, manage server load, bandwidth, database performance, browser and device.

3.3 Software Developer

Software developers are computer programmers that focus on designing and managing programming functions. A function is a section of organized, reusable code that is used to perform an action (functions can also be called methods, subroutines, and procedures).

It is possible for software developers to build entire software applications with only functions. These functions, for example, can enable a person to open their tax file, edit their tax documents, and then save or print them. Each type of program is designed differently and has instructions and specific tasks relevant to the company it services. So trying to edit photos in your bank software won't work because the software developer's design doesn't include any instructions for your bank program to edit photos. Software developers also develop, design, create, and modify programs that run the operating systems for computers, networks and even smartphones. In a nutshell, a software developer's goal is to optimize operational efficiency by designing customized software.

Careers related to software developer: database designer, database developer, game developer, video game engineer, information architect, information systems analyst, information technology analyst (It analyst), information technology consultant (It consultant), interface designer, software analyst, software applications architect, software applications designer, software applications developer, software applications engineer, software applications specialist, software design engineer, software designer, software development engineer, software systems engineer, systems analyst programmer, usability engineer, user interface designer, software application programmer.

Various job tasks for software developers: consult with customers about software system design, review current systems, design software applications, analyze project data to determine specifications or requirements, determine system performance standards, modify existing software to correct errors or to improve its performance, present ideas for system improvements, including cost proposals, work closely with analysts, engineers, programmers, designers and staff, develop and direct software system testing and validation procedures, produce detailed specifications and write the program codes, test the product in controlled, real situations before going live, prepare training manuals for users, maintain the systems once they are up and running, analyze user needs and software requirements, supervise the work of programmers, technologists and technicians.

3.4 Database Developers

A database collects, arranges, sorts and retrieves related pieces of information. It runs behind the scenes of user software and websites, and is generally stored and accessed electronically from a computer system.

Database developers (or database programmers) are the people who are responsible for creating and implementing computer databases. These types of computer programmers will analyze the data needs of a company and then produce an effective database system to meet those needs. They will also test database programs for efficiency and performance and troubleshoot and correct any problems that come up.

Careers related to database developer: database administrator (DBA), network administrator, data architect, database design analyst, database analyst, database coordinator, database design analyst, database designer, database engineer, database management system specialist (dbms specialist), database manager, database modeler, database programmer.

Various job tasks for database developers: design and develop database programs, create databases to store electronic data, work as part of a project team to coordinate database development, develop data model describing data elements and how they are used, analyze existing databases and data needs of clients to develop effective systems, use specific programming languages and codes, follow implementation processes for new databases, troubleshoot and provide solutions for any bugs in new database applications, keep up with new and emerging technologies, use SQL skills, test programs or databases and make necessary modifications, update computer database information.

New Words

programmer	[ˈprəʊɡræmə]	*n.*程序设计者，程序员
coder	['kəʊdə]	*n.*编码员，编码者
specialist	[ˈspeʃəlɪst]	*n.*专家；行家
developer	[dɪˈveləpə]	*n.*开发者
analyst	[ˈænəlɪst]	*n.*分析师
oversimplification	[ˌəʊvəˌsɪmplɪfɪ'keɪʃn]	*n.*过度简化，过分简单化
extraordinary	[ɪkˈstrɔːdɪnərɪ]	*adj.*非凡的；特别的
navigate	[ˈnævɪɡeɪt]	*vt.*驾驶，航行于；使通过
engineer	[ˌendʒɪˈnɪə]	*n.*工程师
debate	[dɪˈbeɪt]	*n.*讨论；辩论；争论
discipline	[ˈdɪsəplɪn]	*n.*学科；训练
standardized	['stændədaɪzd]	*adj.*标准的
certification	[ˌsɜːtɪfɪˈkeɪʃn]	*n.*证明，鉴定，证书
troubleshooting	['trʌblʃuːtɪŋ]	*n.*发现并修理故障
modification	[ˌmɒdɪfɪˈkeɪʃn]	*n.*修改，修正，变更
behaviour	[bɪˈheɪvjə]	*n.*行为；举止
environment	[ɪnˈvaɪrənmənt]	*n.*工作平台；（运行）环境
customize	[ˈkʌstəmaɪz]	*vt.*定制，定做
consistent	[kənˈsɪstənt]	*adj.*一致的；连续的
eliminate	[ɪˈlɪmɪneɪt]	*vt.*排除，消除
routine	[ruːˈtiːn]	*n.*常规；例行程序 *adj.*例行的；常规的
detail-oriented	[ˈdiːteɪl-'ɔːrɪəntɪd]	*adj.*细节导向的，关注细节的
impact	[ˈɪmpækt]	*n.*影响；冲击力 *vt.*对……产生影响 *vi.*产生影响

rectify	[ˈrektɪfaɪ]	*vt.*改正，校正
discrepancy	[dɪsˈkrepənsɪ]	*n.*差异；不符合；不一致
prototype	[ˈprəʊtətaɪp]	*n.*原型，雏形，蓝本
compliance	[kəmˈplaɪəns]	*n.*服从，听从
recommend	[ˌrekəˈmend]	*v.*推荐，建议
appropriate	[əˈprəʊprɪət]	*adj.*适当的；恰当的；合适的
architect	[ˈɑːkɪtekt]	*n.*设计师；创造者
consultant	[kənˈsʌltənt]	*n.*顾问
telecommunication	[ˌtelɪkəˌmjuːnɪˈkeɪʃn]	*n.*电信；通信
dictate	[dɪkˈteɪt]	*vt.*控制，支配
applicable	[əˈplɪkəbl]	*adj.*适当的；可应用的
blog	[blɒg]	*n.*博客
capacity	[kəˈpæsɪtɪ]	*n.*容量；性能；生产能力
experience	[ɪkˈspɪərɪəns]	*n.*体验，经验
register	[ˈredʒɪstə]	*n.&v.*登记，注册
tech	[tek]	*n.*技术
valid	[ˈvælɪd]	*adj.*有效的
practice	[ˈpræktɪs]	*n.*实践，做法；练习
template	[ˈtempleɪt]	*n.*样板；模板
organized	[ˈɔːgənaɪzd]	*adj.*有组织的，有条理的
reusable	[ˌriːˈjuːzəbl]	*adj.*可再用的，可重用的
subroutine	[ˈsʌbruːtiːn]	*n.*子程序
save	[seɪv]	*v.*保存
usability	[ˌjuːzəˈbɪlɪtɪ]	*n.*可用性，适用性
proposal	[prəˈpəʊzl]	*n.*提议，建议
manual	[ˈmænjʊəl]	*n.*手册，指南
supervise	[ˈsuːpəvaɪz]	*v.*监督，管理，指导
arrange	[əˈreɪndʒ]	*v.*安排；整理；排列
coordinate	[kəʊˈɔːdɪneɪt]	*v.*协调，协同
client	[ˈklaɪənt]	*n.*客户；客户端
emerge	[ɪˈmɜːdʒ]	*vi.*出现，浮现

Phrases

software engineer	软件工程师
technological progress	技术进步

social media	社交媒体
concentrate on	专心于，把思想集中于
code walk-through	代码走查
code generation	代码生成
a collection of ...	……的集合
turn on	打开，启动
field service	现场服务
front end	前端
web page	网页
document style guideline	文档样式指南
search engine	搜索引擎
industry standard	行业标准
marketing strategy	营销策略
website map	网站地图
a section of	一段
real situation	实际情况，真实状况
behind the scenes of	在后台，在幕后
come up	发生
data model	数据模型

Abbreviations

DBA (DataBase Administrator)	数据库管理员

Text B 参考译文

计算机程序员

1. 什么是计算机程序员

计算机程序员或编码员是编写计算机软件的人。术语计算机程序员可以指计算机编程的一个领域的专家，也可以指为多种软件编写代码的通才。

术语程序员可以用来指软件开发人员、软件工程师、计算机科学家或软件分析师。但是，这些专业的成员通常具有编程以外的其他软件工程技能。因此，有时将程序员一词视为这些其他职业的极简称呼。

2. 计算机程序员做什么

21 世纪带来了非凡的技术进步。处于这种现代技术中心的是计算机程序员，他们具有创建和引导可能会出现的任何新项目的技术技能。计算机程序员的工作是接受由软件开发人员和工程师创建的设计，并将其转变为计算机可以遵循的指令集。这些指令产生了社交媒体平台、文字处理程序、浏览器以及更多人们每天使用的内容。

关于程序编写在多大程度上是艺术、手工艺或工程学科的争论一直在进行。总的来说，良好的编程被认为是这三者的适度应用，目的是产生一种有效的软件解决方案。

该学科与许多其他技术专业的不同之处在于，计算机程序员通常无须获得执照或通过任何标准化的（或政府监管的）认证测试即可称自己为“程序员”或“软件工程师”。

计算机程序员制定设计、编写、测试、调试/故障排除和维护计算机程序源代码的过程。此源代码是用编程语言编写的，因此计算机可以“理解”它。该代码可能是对现有源代码的修改，也可能是全新的代码。

编程的目的是创建一个产生某一所需行为（客户化）的程序。编写源代码的过程通常需要许多不同学科的专业知识，包括应用程序领域知识、专用算法和形式逻辑。

计算机程序员还设计了一个图形用户界面（GUI），以便非技术用户可以通过简单地单击菜单选项使用该软件。GUI 充当用户和软件代码之间的翻译器。

有些程序员，特别是在从事有许多计算机程序员参与的大型项目时，会使用计算机辅助软件工程（CASE）工具来自动化大部分编码过程。这些工具使程序员能够专注于编写程序的独特部分。从事较小项目的程序员通常会使用“程序员环境”，或者使用整合编译、代码走查、代码生成、测试数据生成和调试功能来提高生产率的应用程序。

计算机程序员还将使用可以针对特定应用程序修改或定制的基本代码库。这种方法产生了更加可靠和一致的程序，并通过消除了一些常规步骤提高了程序员的生产率。计算机程序员还将负责维护程序的运行状况。

随着软件设计的不断发展和某些编程功能的自动化，计算机程序员已开始承担某些以前仅由软件工程师执行的职责。因此，一些计算机程序员现在可以协助软件工程师确定用户需求并设计计算机程序的某些部分。

3. 计算机程序员的类型

如果没有计算机程序员，我们每天使用的产品（例如计算机、智能手机和汽车）将无法执行我们要求他们执行的操作。计算机编程是一个非常注重细节的职业。因此，程序员需要长时间专注于代码，而不能分心或不跟踪进度。通常，很小但很关键的代码问题在技术上可能会产生重大影响，并可能导致程序无法正常运行。尽快发现并纠正小差异的能力对于解决问题至关重要。

3.1 计算机硬件程序员

计算机具有自己的机器语言，可以被计算机理解并从中获取指令。计算机硬件程序员使用特定的机器语言（代码）编写这些指令，以便当有人按下电源按钮，用键盘输入或使用鼠标时计算机知道如何执行。他们还编写代码，以便在计算机打开时可以显示文本。将计算机程序（指令集）放入永久存储器中，以便一旦有人打开计算机，就会显示已编程的信息，提示用户选择计算机接下来要做什么。

计算机硬件程序员还参与研究、设计、开发和测试计算机设备。

计算机硬件程序员的各种工作任务：编写硬件开发过程的详细功能规范；使用模型构建、测试和修改产品原型；设计、分析和测试电气、电子和计算机设备的性能；评估硬件和软件之间的接口；评估操作和绩效要求；准备设计、确定规格并确定操作计划；设计和开发计算机硬件（CPU、支持逻辑、微处理器、电路、打印机和磁盘驱动器）；监控功能并进行必要的修改；监控流程是否符合标准；推荐技术设计或过程变更以提高性能；存储、检索和处理数据以进行分析；分析用户需求并推荐合适的硬件。

与计算机硬件程序员有关的职业：自动化工程师、计算机架构师、计算机工程师、计算机硬件设计师、计算机硬件开发人员、计算机安装工程师、电子工程师、现场服务工程师、硬件设计工程师、硬件工程师、信息技术顾问（IT 顾问）、网络工程师、系统工程师、系统集成工程师、通信工程师。

3.2 网络开发人员

设计、创建和修改因特网上数百万个网站的计算机程序员称为Web开发人员。此类计算机程序员使用的软件可以让他们决定人们访问网站时可以执行哪些功能。

Web开发人员首先要在设计和构建网站之前分析用户的需求。如果需要，他们还会添加适用的图形、音频和视频组件。即使是简单的博客也需要Web开发人员来设计观众可以看到的结构、功能和信息。

Web开发人员不仅要负责网站的外观，而且还要负责网站的性能、容量，有时也创建内容。网站的常见编程语言包括超文本标记语言、JavaScript、Drupal、WordPress和Joomla。

与Web开发人员有关的职业：前端Web开发人员、互联网架构师、PHP Web开发人员、可用性专家、用户体验设计师、用户界面开发人员、Web应用程序开发人员、网站架构师、网页开发人员、网络程序员、网站开发人员、网站专员、网络专家、网站管理员。

Web开发人员的各种工作任务：分析用户需求，选择编程语言、设计工具或应用程序；创建网络模型或原型；编写计算机程序代码；设计、建立、编辑或维护网站；使用内容创建工具、管理工具和数字媒体；制定并记录网站内容的文档样式指南；执行或指导网站更新；用搜索引擎注册网站；优先考虑需求，解决技术问题并制定内容标准；将文件备份到本地目录以防止信息丢失；解决软件问题并排除故障；确保代码有效并符合行业标准；确保代码与浏览器、设备或操作系统兼容；更新有关当前 Web 技术和实践的知识；制定测试计划以测试性能；创建和开发支持 Web 应用程序和网站的数据库；制定和整合电子商务战略和营销战略；为计算机网络问题提供技术支持；设计和实施防火墙或消息加密；开发网站地图、应用程序模型、图像模板或页面模板；准备信息的图形或其他视觉表示；管理与其他网站之间的链接；管理服务器负载、带宽、数据库性能、浏览器和设备。

3.3 软件开发人员

软件开发人员是专注于设计和管理编程函数的计算机程序员。函数是用于执行操作的有组织的、可重用代码的一部分（函数也可以称为方法、子例程和过程）。

软件开发人员可以仅使用函数来构建整个软件应用程序。例如，这些函数可以使一个人打开其税务文件，编辑其税务文件，然后保存或打印它们。每种程序的设计都不同，并具有与其所服务的公司有关的指令和特定任务。因此，尝试在银行软件中编辑照片是行不通的，因为软件开发人员的设计并未包含银行软件编辑照片的任何指令。软件开发人员还开发、设计、创建和修改运行计算机、网络甚至智能手机的操作系统的程序。简而言之，软件开发人员的目标是通过设计定制软件来优化操作效率。

与软件开发人员有关的职业：数据库设计师、数据库开发人员、游戏开发者、电子游戏工程师、信息架构师、信息系统分析师、信息技术分析师（IT 分析师）、信息技术顾问（IT 顾问）、界面设计师、软件分析师、软件应用架构师、软件应用程序设计师、软件应用程序开发人员、软件应用工程师、软件应用专家、软件设计工程师、软件设计师、软件开发工程师、软件系统工程师、系统分析程序员、可用性工程师、用户界面设计器、软件应用程序员。

针对软件开发人员的各种工作任务：与客户协商软件系统设计；审查当前系统；设计软件应用程序；分析项目数据以确定规格或要求；确定系统性能标准；修改现有软件以纠正错误或提高其性能；提出系统改进的想法，包括费用建议；与分析师、工程师、程序员、设计师和员工紧密合作；制定并指导软件系统测试和确认程序；制定详细的规格并编写程序代码；在投入使用之前，在受控的真实环境中测试产品；为用户准备培训手册；一旦系统启动并运行，对其进行维护；分析用户需求和软件要求；监督程序员、技术人员和技术员的工作。

3.4 数据库开发人员

数据库收集、整理、分类和检索相关的信息。它运行在用户软件和网站的幕后，通常从计算机系统进行电子存储和访问。

数据库开发人员（或数据库程序员）是负责创建和实现计算机数据库的人员。这些类型的计算机程序员将分析公司的数据需求，然后制作一个满足这些需求的有效的数据库系统。他们还将测试数据库程序的效率和性能，并对所有出现的问题进行故障排除和纠正。

与数据库开发人员有关的职业：数据库管理员（DBA）、网络管理员、数据架构师、数据库设计分析师、数据库分析师、数据库协调员、数据库设计分析师、数据库设计员、数据库工程师、数据库管理系统专家（DBMS 专家）、数据库管理员、数据库建模员、数据库程序员。

数据库开发人员的各种工作任务：设计和开发数据库程序；创建数据库以存储电子数据；作为项目团队的一部分来协调数据库开发；开发描述数据元素及其使用方式的数据模型；分析客户现有的数据库和数据需求以开发有效的系统；使用特定的编程语言和代码；从事新数据库的实施过程；对新数据库应用程序中的任何错误进行故障排除并提供解决方案；跟上新兴技术；使用 SQL 技巧；测试程序或数据库并进行必要的修改；更新计算机数据库信息。

Exercises

[Ex. 1] Answer the following questions according to Text A.

1. What is object–oriented analysis (OOA)?
2. What does the implementation details of object–oriented design (OOD) generally include?
3. What is object-oriented programming (OOP)?
4. What is an object? What does it represent?
5. What is encapsulation? What is data hiding or information hiding ?
6. What are the features of message passing?
7. What are the types of inheritance?
8. What does polymorphism imply in object-oriented paradigm?
9. What do generalization and specialization represent?

[Ex.2] Fill in the following blanks according to Text B.

1. A computer programmer, or coder, is someone who ____________________. The term programmer can be used to refer to a software developer, ______________, ______________, or ______________.

2. It's the job of ________________ to take designs created by software developers and engineers and ________________ that computers can follow.

3. The discipline differs from many other technical professions in that a computer programmer, in general, ________________ or ________________ in order to call themselves a "programmer" or even a "software engineer."

4. 4.The process of writing source code often requires expertise in many different subjects, including ____________________, ________________, and ____________.

5. A programmer working on smaller projects will often use “programmer environments”, or applications that increase productivity by ____________________, ________________, ____________________, ____________________ and ____________________.

6.Computer hardware programmers write these instructions in a specific machine language (code) so that a computer knows what to do when someone ________, ________, or ________. They also write code so that text can be displayed ________.

7. Web developers start by ________ before designing and structuring a website. They also add ________, ________, and ________ if needed.

8. Common programming languages for websites include ________, ________, Drupal, ________, and ________.

9. Software developers also develop, design, create, and ________ programs that run the operating systems for ________, ________ and even ________.

10. Database developers are the people who are responsible for ________ and ________ computer databases. These types of computer programmers will analyze ________ of a company and then produce an effective ________ to meet those needs.

[Ex. 3] Translate the following terms or phrases from English into Chinese and vice versa.

1. analysis model	1. ________
2. hybrid inheritance	2. ________
3. message passing	3. ________
4. object model diagram	4. ________
5. object-oriented technology	5. ________
6. *n.*属性，性质，特征	6. ________
7. *n.*类	7. ________
8. *n.*继承；遗传	8. ________
9. *n.*实例化	9. ________
10. *n.*对象；物体；目标；客体	10. ________

[Ex. 4] Translate the following sentences into Chinese.

Computer Programmer

A computer programmer is a skilled professional who codes, tests, debugs, and maintains the comprehensive instructions known as computer programs that devices should follow to execute their functions.

Computer programmers also conceptualize, design, and test logical structures to solve computer issues. Programmers make use of specific computer languages like C, C++, Java, PHP, .NET, etc. to convert the program designs developed by software developers or system architects into instructions that the computer could follow. They often refer to code libraries for simplifying their coding, and might build or make use of computer-aided software tools to automate the coding.

A computer programmer is also referred to as a programmer, coder, developer, or software engineer. Also, the term is often used to refer to a stand-alone software developer, mobile applications developer, web developer, software analyst, embedded firmware developer, and so on.

Computer programmers are generally classified into two kinds: application programmers and systems programmers.

Application programmers perform coding to manage a certain task, such as coding a program to monitor inventory within a company. On the other hand, systems programmers code programs to maintain and control system software, including database management systems and operating systems (OSs).

Programmers utilize programming editors, also referred to as source-code editors, to write the source code of a program or an application. These types of editors incorporate features ideal for programmers, which includes color-syntax highlighting, auto-complete, auto indentation, syntax check, bracket matching, etc. These features help the programmers throughout coding, debugging and testing.

[Ex. 5] Fill in the blanks with the words given below.

compiler	modification	executable	interpretation	interpreted
convert	machine	translates	statement	changes

Compiler and Interpreter

1. What Is Compiler

A compiler is a computer program that transforms code written in a high-level programming language into the machine code. It is a program which ___1___ the human-readable code to a language a computer processor understands (binary 1 and 0 bits). The computer processes the ___2___ code to perform the corresponding tasks.

A compiler should comply with the syntax rule of that programming language in which it is written. However, the ___3___ is only a program and cannot fix errors found in that program. So, if you make a mistake, you need to make ___4___ in the syntax of your program. Otherwise, it will not compile.

A compiler works with what are sometimes called 3GL and higher-level languages. It reads the source code and outputs ___5___ code. It translates software written in a higher-level language into instructions that computer can understand. It converts the text that a programmer writes into a format the CPU can understand.

2. What Is Interpreter

An interpreter is a computer program, which coverts each high-level program ___6___ into the machine code. This includes source code, pre-compiled code, and scripts. Both compiler and interpreters do the same job, which is converting higher–level programming language to machine code. However, a compiler will convert the code into machine code (create an exe) before the program is run. Interpreters ___7___ code into machine code when the program is run.

The interpreter converts the source code line-by-line during RUN time. It completely translates a program written in a high-level language into machine–level language. It allows evaluation and ___8___ of the program while it is executing.

3. Key Differences

- A compiler transforms code written in a high-level programming language into the machine code, at once, before the program is run, whereas an interpreter coverts each high-level program statement, one by one, into the machine code, during the program is run.
- Compiled code runs faster while ___9___ code runs slower.
- The compiler displays all errors after compilation while the interpreter displays errors of each line one by one.
- Compiler is based on translation linking-loading model, whereas interpreter is based on ___10___ method.
- The compiler takes an entire program whereas the interpreter takes a single line of code.

Online Resources

二维码	内　容
	计算机专业常用语法（6）：动名词
	在线阅读（1）：4 Advantages of Object-Oriented Programming （面向对象编程的 4 个优点）
	在线阅读（2）：The Differences Between HTML and HTML5 （HTML 与 HTML5 的不同之处）

Unit 7

Programming Languages

Text A

Programming Language

扫码听课文

Programming language is the language of computers. Through programming language, we can communicate with a computer system. Computers can only understand binary, but humans are not comfortable with binary number system. Humans cannot interact fluently with computers in the language of 0's and 1's. Programming language act as an interface between computers and humans.

Programming languages are used to create programs. A computer program is intended to perform some specific tasks through computer or to control the behavior of computer.

Using a programming language, we write instructions that the computer should perform. Instructions are usually written using characters, words, symbols and decimal. These instructions are later encoded to the computer understandable language i.e. binary language, so that the computer can understand the instructions given by human and can perform specified task.

Today thousands of programming language have been created and many are still being developed every year. Every programming language is designed for some specific purposes. Such as FORTRAN, OCaml, Haskell are best suited for scientific and numerical computations, whereas Java, C++, C# are best suited for designing server applications, games, desktop applications and many more.

Programming languages are basically classified into two main categories – low-level language and high-level language. However, there also exists another category known as middle level language. Every programming language belongs to one of these categories and subcategories.

1. Low-Level Languages

Low-level languages are languages close to the machine-level instruction set. They provide less or no abstraction from the hardware. A low-level programming language interacts directly with the registers and memory. Instructions written in low-level languages are machine-dependent.

Programs developed using low-level languages are machine-dependent and are not portable.

Low-level language does not require any compiler or interpreter to translate the source code into machine code. An assembler may translate the source code written in low-level language into machine code.

Programs written in low-level languages are fast and memory-efficient. However, it is nightmare for programmers to write, debug and maintain low-level programs. They are mostly used to develop operating systems, device drivers and applications that require direct hardware access.

Low-level languages are further classified into two more categories: machine language and assembly language.

1.1 Machine Language

Machine language is the closest language to the hardware. It consists of a set of instructions that are executed directly by the computer. These instructions are a sequence of binary bits. Each instruction performs a very specific and small task. Instructions written in machine language are machine-dependent and varies from computer to computer.

A programmer must have additional knowledge about the architecture of the particular machine before programming in machine language. Developing programs using machine language is a tedious job. It is very difficult to remember sequence of binaries for different computer architectures. Therefore, nowadays it is not much in practice.

1.2 Assembly Language

Assembly language is an improvement over machine language. Similar to machine language, assembly language also interacts directly with the hardware. Instead of using raw binary sequence to represent an instruction set, assembly language uses mnemonics.

Mnemonics are short abbreviated English words used to specify a computer instruction. Each instruction in binary has a specific mnemonic. They are architecture-dependent and there is a list of separate mnemonics for different computer architectures.

Assembly language uses a special program called assembler. Assembler translates mnemonics into specific machine code.

Assembly language is still in use. It is used for developing operating systems, device drivers, compilers and other programs that requires direct hardware access.

1.3 Advantages of Low Level Languages

- Programs developed using low-level languages are fast and memory efficient.
- Programmers can utilize processor and memory in a better way.
- There is no need of any compiler or interpreter to translate the source code into machine code, thus cutting the compilation and interpretation time.
- Low-level languages provide direct manipulation of computer registers and storage.
- It can directly communicate with hardware devices.

1.4 Disadvantages of Low Level Languages

- Programs developed using low-level languages are machine-dependent and are not portable.
- It is difficult to develop, debug and maintain.
- Low-level programs are more error prone.
- Low-level programming usually results in poor programming productivity.
- Programmer must have additional knowledge of the computer architecture of particular machine.

2. High-Level Languages

High-level languages are similar to the human language. Unlike low-level languages, high-level languages are programmer-friendly, easy to code, debug and maintain.

High-level language provides higher level of abstraction from machine language. They do not interact directly with the hardware. Rather, they focus more on the complex arithmetic operations, optimal program efficiency and easiness in coding.

High-level programs require compilers/interpreters to translate source code into machine language. We can compile the source code written in high-level language to multiple machine languages. Thus, they are machine-independent language.

Today almost all programs are developed using a high-level programming language. We can develop a variety of applications using high-level language. They are used to develop desktop applications, websites, system software, utility software and many more.

High-level languages are grouped in two categories based on execution model: compiled or interpreted languages.

On the basis of programming paradigm, high-level languages can also be classified into structured programming language, procedural programming language, and object-oriented programming language.

2.1 Advantages of High-Level Language

- High-level languages are programmer friendly. They are easy to write, debug and maintain.
- It provides higher level of abstraction from machine languages.
- It is a machine-independent language.
- Easy to learn.
- Less error prone, easy to find and debug errors.
- High-level programming results in better programming productivity.

2.2 Disadvantages of High-Level Language

- It takes additional translation times to translate the source code into machine code.
- High-level programs are comparatively slower than low-level programs.

- Compared to low-level programs, they are generally less memory-efficient.
- It cannot communicate directly with the hardware.

3. Differences Between Low-Level Language and High-Level Language

3.1 Program Speed

Programs in low-level language are written either in binary or assembly language. They do not require any compilation or interpretation. It interact directly with the registers and memory. Thus, they are comparatively faster than high-level languages.

High-level language uses English statements to write programs. Hence, they require compilers or interpreters to translate the source code into machine language. They do not interact directly with the hardware. Thus, they are slower than low-level languages.

3.2 Memory Efficiency

Low-level languages are memory-efficient. They generally consume less memory.

High-level languages are not memory-efficient. They generally run inside a specific runtime environment. Also there are several other programs running concurrently to increase optimal efficiency of the program which consumes memory. Thus, the overall memory consumption of high-level language is comparatively more than low-level language.

3.3 Easiness

Low-level languages are machine-friendly languages. To write a program in low-level language, we must know binaries or mnemonics of low-level instruction sets. Remembering various instructions sets for different architectures is nearly impossible. Thus, low-level programming is difficult to learn. Learning low-level languages requires additional knowledge and experience about the specific machine architecture.

High-level languages are programmer-friendly languages. Programs in high-level language are written using English statements, which is much easier to remember than low-level binaries or mnemonics. Hence, high-level programming is easy to learn.

3.4 Portability

Low-level language contain low-level computer instructions set. These instructions are machine-dependent and are different for different architectures. Hence, programs developed are also machine-dependent and are not portable.

High-level languages uses English statements to write programs. They are further translated into machine language using a compiler or interpreter. There exists a separate compiler or interpreter for different machine architectures. That translates the source code into specific machine language. Hence, high-level languages are machine-independent and are portable.

3.5 Abstraction Level

Low-level language provides less or no abstraction from the hardware. They are the closest language to the hardware. They interact directly with the computers register and memory.

High-level language provides a high level of abstraction from the hardware. They run on top of the machine language. They do not interact directly with the computers register and memory. They interact with the hardware through operating system.

3.6 Debugging and Maintenance

Low-level languages are more error-prone, from small syntactical error to big memory leaks. Error detection and maintenance is a tedious and time-consuming process.

High-level languages are less error-prone. Almost all syntactical errors are identified using compilers or interpreters. They are generally easy to debug and maintain.

3.7 Additional Knowledge and Experience

Low-level languages are machine-dependent. Programmer require a prior knowledge of the particular computer architecture before one can actually write a program for that computer.

High-level languages are machine-independent. Programmer do not require any prior knowledge of the computer architecture.

3.8 Applications

Low-level languages interacts directly with the hardware. They provide very less or no abstraction from the hardware. They are generally used to develop operating systems and embedded systems.

High-level languages provide a higher level of abstraction from the hardware. Nowadays, almost all software are developed using a high-level language. It is used to develop variety of applications such as desktop applications, websites, utility software, mobile applications etc.

New Words

binary	[ˈbaɪnərɪ]	*adj.*二进制的
fluently	['flu:əntlɪ]	*adv.*流利地，流畅地
decimal	[ˈdesɪml]	*adj.*十进位的，小数的 *n.*小数
category	[ˈkætəgərɪ]	*n.*种类，类别，类目
abstraction	[æbˈstrækʃn]	*n.*抽象；抽象概念；抽象化
dependent	[dɪˈpendənt]	*adj.*依靠的，依赖的
portable	[ˈpɔ:təbl]	*adj.*轻便的
nightmare	[ˈnaɪtmeə]	*n.*噩梦；可怕的事情
tedious	[ˈti:dɪəs]	*adj.*单调沉闷的；令人生厌的
mnemonic	[ni:'mɒnɪk]	*n.*助记符
utilize	[ˈju:təlaɪz]	*vt.*利用，使用

prone	[prəʊn]	*adj.*易于……的；有……倾向的
arithmetic	[əˈrɪθmətɪk]	*n.*计算，算法
compile	[kəmˈpaɪl]	*vt.*编译
execution	[ˌeksɪˈkju:ʃn]	*n.*执行，实行
comparatively	[kəmˈpærətɪvlɪ]	*adv.*对比地，相对地，比较地
compilation	[ˌkɒmpɪˈleɪʃn]	*n.*编译
interpretation	[ɪnˌtɜ:prɪˈteɪʃn]	*n.*解释，说明
consume	[kənˈsju:m]	*vt.*消耗，消费
runtime	[rʌn'taɪm]	*n.*运行时间，运行期
concurrently	[kən'kʌrəntlɪ]	*adv.*同时地
easiness	['i:zɪnɪs]	*n.*简易性
portability	[ˌpɔ:tə'bɪlɪtɪ]	*n.*便携性
contain	[kənˈteɪn]	*vt.*包含，容纳；包括
syntactical	[sɪn'tæktɪkəl]	*adj.*依照句法的
mobile	[ˈməʊbaɪl]	*adj.*可移动的

Phrases

programming language	编程语言
communicate with ...	与……通信
be comfortable with	和某人相处自在
interact with ...	与……交互，与……相互影响，与……相互配合
be intended to	打算，意图是
numerical computation	数字计算
be classified into	分类为
high-level language	高级语言
middle-level language	中级语言
instruction set	指令集
assembly language	汇编语言
utility software	实用软件
structured programming language	结构化编程语言
procedural programming language	过程编程语言
object-oriented programming language	面向对象编程语言
memory leak	内存泄漏
embedded system	嵌入式系统

Text A 参考译文

编 程 语 言

编程语言是计算机的语言。通过编程语言，我们可以与计算机系统进行通信。计算机只能理解二进制，但人类不能适应二进制数字系统。人类无法以 0 和 1 的语言流畅地与计算机交互。编程语言充当了计算机与人之间的接口。

编程语言用于创建程序。计算机程序旨在通过计算机执行某些特定任务或控制计算机的行为。

我们使用编程语言编写计算机应执行的指令。指令通常使用字符、单词、符号和十进制数来编写。这些指令随后被编码为计算机可理解的语言，即二进制语言，以便计算机可以理解由人给出的指令并可以执行指定的任务。

如今，已经创建了数千种编程语言，并且每年仍在开发中。每种编程语言都是为特定目的而设计的。例如 FORTRAN、OCaml、Haskell 最适合科学和数值计算，而 Java、C ++、C # 最适合设计服务器应用程序、游戏、桌面应用程序等。

编程语言基本上分为两大类——低级语言和高级语言。但是，也存在另一种称为中级语言的类别。每种编程语言都属于这些类别和子类别之一。

1. 低级语言

低级语言是接近机器级指令集的语言。它们很少提供或不提供硬件抽象。低级编程语言直接与寄存器和存储器交互。用低级语言编写的指令依赖机器。使用低级语言开发的程序依赖计算机，并且不可移植。

低级语言不需要任何编译器或解释器就可将源代码转换为机器代码。汇编器可以把用低级语言编写的源代码转换为机器代码。

用低级语言编写的程序速度快且存储效率高。但是，编写、调试和维护低级程序是程序员的噩梦。它们主要用于开发需要直接访问硬件的操作系统、设备驱动程序和应用程序。

低级语言又分为两类：机器语言和汇编语言。

1.1 机器语言

机器语言是最接近硬件的语言。它由计算机直接执行的一组指令组成。这些指令是二进制位的一个序列。每个指令执行一个非常具体的小任务。用机器语言编写的指令依赖机器，并且因计算机而异。

在使用机器语言进行编程之前，程序员必须具备特定机器体系结构的相关知识。使用机器语言开发程序是一项烦琐的工作。记住不同计算机体系结构的二进制序列非常困难。因此，如今在实践中机器语言已经不常用了。

1.2 汇编语言

汇编语言是对机器语言的改进。与机器语言类似，汇编语言也直接与硬件交互。汇编语言不是使用原始的二进制序列来表示指令集，而是使用助记符。

助记符是英文缩写词，用于指定计算机指令。每个二进制指令都有一个特定的助记符。它们依赖于体系结构，并且为不同的计算机体系结构提供了不同的助记符列表。

汇编语言使用一种称为汇编程序的特殊程序。汇编程序将助记符转换为特定的机器代码。

汇编语言仍在使用。它用于开发需要直接访问硬件的操作系统、设备驱动程序、编译器和其他程序。

1.3 低级语言的优势

- 使用低级语言开发的程序快速且存储效率高。
- 程序员可以更好地利用处理器和内存。
- 无须任何编译器或解释器即可将源代码转换为机器代码，从而减少了编译和解释时间。
- 低级语言可直接操纵计算机寄存器和存储。
- 它可以直接与硬件设备通信。

1.4 低级语言的缺点

- 使用低级语言开发的程序依赖于计算机，并且不可移植。
- 难以开发、调试和维护。
- 低级程序更容易出错。
- 低级编程通常会导致编程效率低下。
- 程序员必须对特定机器的计算机体系结构有更多的了解。

2. 高级语言

高级语言与人类语言相似。与低级语言不同，高级语言对程序员友好，易于编码、调试和维护。

高级语言提供了对机器语言的更高层次的抽象。它们不直接与硬件交互。相反，它们更多地关注复杂的算术运算、优化程序效率和编码简便性。

高级程序要求编译器/解释器将源代码转换为机器语言。我们可以把用高级语言编写的源代码编译为多种机器语言。因此，它们是独立于机器的语言。

如今，几乎所有程序都是使用高级编程语言开发的。我们可以使用高级语言开发各种应用程序。它们用于开发桌面应用程序、网站、系统软件、实用程序软件等。

根据执行模型，高级语言分为两类：编译语言或解释语言。

也可以根据编程范例将高级语言分类为结构化编程语言、过程编程语言和面向对象编程语言。

2.1 高级语言的优势

- 高级语言对程序员友好。它们易于编写、调试和维护。
- 它提供了更高级别的机器语言抽象。
- 这是一种独立于机器的语言。
- 易于学习。
- 不易犯错，易于发现和调试错误。
- 高级编程可提高编程效率。

2.2 高级语言的缺点

- 将源代码转换为机器代码需要花费额外的转换时间。
- 高级程序比低级程序要慢。
- 与低级程序相比，它们的内存效率通常较低。
- 它无法直接与硬件通信。

3. 低级语言和高级语言之间的差异

3.1 程序速度

低级语言的程序以二进制或汇编语言编写。它们不需要任何编译或解释。它直接与寄存器和存储器交互。因此，它们比高级语言要快。

高级语言使用英语语句编写程序。因此，它们需要编译器或解释器将源代码转换为机器语言。它们不直接与硬件交互。因此，它们比低级语言慢。

3.2 存储效率

低级语言存储效率高。它们通常消耗较少的内存。

高级语言存储效率不高。它们通常在特定的运行时环境中运行。另外，还有其他几个程序同时运行以提高程序的最佳效率，但消耗内存。因此，高级语言的总体内存消耗相对比低级语言要多。

3.3 简易性

低级语言是机器友好的语言。要使用低级语言编写程序，我们必须了解低级指令集的二进制或助记符。记住针对不同体系结构的各种指令集几乎是不可能的。因此，低级编程很难学习。学习低级语言需要有关特定机器体系结构的额外知识和经验。

高级语言是程序员友好的语言。用高级语言编写的程序是使用英语语句编写的，这比低级二进制或助记符更容易记住。因此，高级编程易于学习。

3.4 便携性

低级语言包含低级计算机指令集。这些指令依赖于机器，并且对于不同的体系结构是不同的。因此，开发的程序也依赖于机器并且不可移植。

高级语言使用英语语句编写程序。使用编译器或解释器将它们进一步翻译为机器语言。不同的机器体系结构有不同的编译器或解释器。这会将源代码转换成特定的机器语言。因此，高级语言是独立于机器的并且是可移植的。

3.5 抽象级别

低级语言很少或根本没有提供来自硬件的抽象。它们是最接近硬件的语言。它们直接与计算机的寄存器和内存交互。

高级语言提供了硬件的高级抽象。它们在机器语言之上运行。它们不会直接与计算机的寄存器和内存进行交互。它们通过操作系统与硬件交互。

3.6 调试与维护

低级语言更容易出错，从小的语法错误到大的内存泄漏。错误检测和维护是一个烦琐且耗时的过程。

高级语言不太容易出错。编译器或解释器可以识别几乎所有的语法错误。它们通常易于调试和维护。

3.7 其他知识和经验

低级语言依赖于机器。程序员需要先了解特定计算机体系结构的知识，然后才能实际为该计算机编写程序。

高级语言独立于机器。程序员不需要任何计算机体系结构的先备知识。

3.8 应用

低级语言直接与硬件交互。它们很少提供或不提供硬件抽象。它们通常用于开发操作系统和嵌入式系统。

高级语言提供了更高层次的硬件抽象。如今，几乎所有软件都是使用高级语言开发的。它用于开发各种应用程序，例如桌面应用程序、网站、实用程序软件、移动应用程序等。

Text B

Python Programming Language

扫码听课文

Python features a dynamic type system and automatic memory management. It supports multiple programming paradigms, including object-oriented, imperative, functional and procedural, and has a large and comprehensive standard library.

1. Features and Philosophy

Python is a multi-paradigm programming language. Object-oriented programming and structured programming are fully supported, and many of its features support functional programming and aspect-oriented programming. Many other paradigms are supported via extensions, including design by contract and logic programming.

Python uses dynamic typing, and a combination of reference counting and a cycle-detecting garbage collector for memory management. It also features dynamic name resolution (late binding), which binds method and variable names during program execution.

2. Syntax and Semantics

Python is meant to be an easily readable language. Its formatting is visually uncluttered, and it often uses English keywords where other languages use punctuation. Unlike many other languages, it does not use curly brackets to delimit blocks, and semicolons after statements are optional. It has fewer syntactic exceptions and special cases than C or Pascal.

2.1 Indentation

Python uses whitespace indentation, rather than curly brackets or keywords, to delimit blocks. An increase in indentation comes after certain statements; a decrease in indentation signifies the end of the current block. Thus, the program's visual structure accurately represents the program's semantic structure. This feature is also sometimes termed the off-side rule.

2.2 Statements and Control Flow

Python's statements include:

- The assignment statement (token "=", the equals sign).
- The if statement, which conditionally executes a block of code, along with else and elif (a contraction of else-if).
- The for statement, which iterates over an iterable object, capturing each element to a local variable for use by the attached block.
- The while statement, which executes a block of code as long as its condition is true.
- The raise statement, used to raise a specified exception or re-raise a caught exception.
- The class statement, which executes a block of code and attaches its local namespace to a class, for use in object-oriented programming.
- The def statement, which defines a function or method.
- The pass statement, which serves as a NOP. It is syntactically needed to create an empty code block.
- The assert statement, used during debugging to check for conditions that ought to apply.
- The yield statement, which returns a value from a generator function. From Python 2.5, yield is also an operator.
- The import statement, which is used to import modules whose functions or variables can

be used in the current program. There are three ways of using import: import <module name> [as <alias>] or from <module name> import * or from <module name> import <definition 1> [as <alias 1>], <definition 2> [as <alias 2>], ...

- The print statement was changed to the print() function in Python 3.

2.3 Expressions

Some Python expressions are similar to languages such as C and Java, while some are not:

(1) Addition, subtraction, and multiplication are the same, but the behavior of division differs. There are two types of divisions in Python. Python also added the ** operator for exponentiation.

(2) From Python 3.5, the new @ infix operator was introduced. It is intended to be used by libraries such as NumPy for matrix multiplication.

(3) In Python, == compares by value, versus Java, which compares numerics by value and objects by reference. Python's is operator may be used to compare object identities. In Python, comparisons may be chained, for example *a* <= *b* <= *c*.

(4) Python uses the words and, or, not for its boolean operators rather than the symbolic &&, ||, ! used in Java and C.

(5) Python has a type of expression termed a list comprehension. Python 2.4 extended list comprehensions into a more general expression termed a generator expression.

(6) Conditional expressions in Python are written as *x* if *c* else y (different in order of operands from the *c* ? *x* : *y* operator common to many other languages).

(7) Python makes a distinction between lists and tuples. Lists are written as [1, 2, 3], are mutable, and cannot be used as the keys of dictionaries (dictionary keys must be immutable in Python). Tuples are written as (1, 2, 3), are immutable and thus can be used as the keys of dictionaries, provided all elements of the tuple are immutable.

(8) Python has a "string format" operator %. This functions analogous to printf format strings in C, e.g. "spam=%s eggs=%d" % ("blah", 2).

(9) Python has various kinds of string literals:

- Strings delimited by single or double quote marks. Both kinds of string use the backslash (\) as an escape character.
- Triple-quoted strings, which begin and end with a series of three single or double quote marks.
- Raw string, denoted by prefixing the string literal with an r. Escape sequences are not interpreted.

3. Libraries

Python's large standard library, commonly cited as one of its greatest strengths, provides tools suited to many tasks. For Internet-facing applications, many standard formats and protocols such as MIME and HTTP are supported. It includes modules for creating graphical user interfaces, connecting to relational databases, generating pseudorandom numbers, arithmetic with arbitrary precision decimals, manipulating regular expressions, and unit testing.

Some parts of the standard library are covered by specifications, but most modules are not. They are specified by their code, internal documentation, and test suites (if supplied). However, because most of the standard library is cross-platform Python code, only a few modules need altering or rewriting.

As of March 2018, the Python Package Index (PyPI), the official repository for third-party Python software, contains over 130,000 packages with a wide range of functionality, including:

- Graphical user interfaces
- Web frameworks
- Multimedia
- Databases
- Networking
- Test frameworks
- Automation
- Web scraping
- Documentation
- System administration
- Scientific computing
- Text processing
- Image processing

4. Development Environments

Most Python implementations (including CPython) include a read–eval–print loop (REPL), permitting them to function as a command line interpreterfor which the user enters statements sequentially and receives results immediately.

Other shells add further abilities such as auto-completion, session state retention and syntax highlighting.

As well as standard desktop integrated development environments, there are web browser-based IDEs: SageMath, intended for developing science and math-related Python programs; PythonAnywhere, a browser-based IDE and hosting environment; and Canopy IDE, a commercial Python IDE emphasizing scientific computing.

New Words

feature	[ˈfiːtʃə]	*n.*特征，特点
		*vt.*使有特色；描写……的特征
dynamic	[daɪˈnæmɪk]	*adj.*动态的
object-oriented	[ˈɒbdʒɪkt-ˈɔːrɪəntɪd]	*adj.*面向对象的

imperative	[ɪmˈperətɪv]	*adj.*命令的 *n.*命令；规则
functional	[ˈfʌŋkʃənl]	*adj.*功能的；函数的
procedural	[prəˈsi:dʒərəl]	*adj.*程序的，过程的
philosophy	[fəˈlɒsəfɪ]	*n.*哲学；哲学体系
metaprogramming	[metæpˈrəʊgræmɪŋ]	*adj.*元程序设计的，元编程的
metaobject	[metæˈɒbdʒɪkt]	*adj.*元对象的
logic	[ˈlɒdʒɪk]	*n.*逻辑，逻辑学 *adj.*逻辑的
resolution	[ˌrezəˈlu:ʃn]	*n.*解析；分辨率
binding	[ˈbaɪndɪŋ]	*n.*绑定 *adj.*捆绑的 *v.*绑定
variable	[ˈveərɪəbl]	*n.*变量 *adj.*变量的
syntax	[ˈsɪntæks]	*n.*语法
semantic	[sɪˈmæntɪk]	*adj.*语义的
uncluttered	[ˌʌnˈklʌtəd]	*adj.*整齐的，整洁的
punctuation	[ˌpʌŋktʃʊˈeɪʃn]	*n.*标点法；标点符号
semicolon	[ˌsemɪˈkəʊlən]	*n.*分号
statement	[ˈsteɪtmənt]	*n.*声明；语句
optional	[ˈɒpʃənl]	*adj.*可选择的；任意的；非强制的
indentation	[ˌɪndenˈteɪʃn]	*n.*缩进，缩格
delimit	[dɪˈlɪmɪt]	*vt.*界定；限制
accurately	[ˈækjʊrətlɪ]	*adv.*正确无误地，准确地；精确地
represent	[ˌreprɪˈzent]	*v.*代表
conditionally	[kənˈdɪʃənəlɪ]	*adv.*有条件地
execute	[ˈeksɪkju:t]	*vt.*执行
iterate	[ˈɪtəreɪt]	*vt.*迭代，重复
attached	[əˈtætʃt]	*adj.*附加的，附属的
condition	[kənˈdɪʃn]	*n.*条件；状态
exception	[ɪkˈsepʃn]	*n.*异常；例外，除外
namespace	[ˈneɪmspeɪs]	*n.*命名空间，名字空间
syntactically	[sɪnˈtæktɪklɪ]	*adv.*依照句法地，在语句构成上地
create	[krɪˈeɪt]	*vt.*建立，创建

return	[rɪˈtɜːn]	*v.*返回
generator	[ˈdʒenəreɪtə]	*n.*生成器
infix	[ˈɪnfɪks]	*vt.*让…插进 *n.*插入词，中缀
boolean	[ˈbuːlɪən]	*adj.*布尔的
symbolic	[sɪmˈbɒlɪk]	*adj.*符号的；象征性的
comprehension	[ˌkɒmprɪˈhenʃn]	*n.*推导式
operand	[ˈɒpərænd]	*n.*操作数；运算数
tuple	[tʌpl]	*n.*元组
mutable	[ˈmjuːtəbl]	*adj.*可变的，易变的
analogous	[əˈnæləgəs]	*adj.*相似的，可比拟的
backslash	[ˈbækslæʃ]	*n.*反斜线符号
interpret	[ɪnˈtɜːprɪt]	*v.*解释，说明
library	[ˈlɑɪbrərɪ]	*n.*库
standard	[ˈstændəd]	*n.*标准，规格 *adj.*标准的
strength	[streŋθ]	*n.*力量；优点，长处
task	[tɑːsk]	*n.*工作，任务；作业
connect	[kəˈnekt]	*vt.*连接，联结 *vi.*连接；建立关系
pseudorandom	[psjuːdəʊ'rændəm]	*adj.*伪随机的
arbitrary	[ˈɑːbɪtrərɪ]	*adj.*任意的，随意的
cross-platform	[krɔsˈplætfɔːm]	*n.*跨平台
rewrite	[ˌriːˈraɪt]	*vt.*重写，改写
package	[ˈpækɪdʒ]	*n.*软件包
framework	[ˈfreɪmwɜːk]	*n.*构架；框架
networking	[ˈnetwɜːkɪŋ]	*n.*网络化
automation	[ˌɔːtəˈmeɪʃn]	*n.*自动化（技术），自动操作
scrape	[skreɪp]	*v.*抓取
session	[ˈseʃn]	*n.*会话
retention	[rɪˈtenʃn]	*n.*保留
emphasize	[ˈemfəsaɪz]	*vt.*强调，着重；使突出

Phrases

automatic memory management	自动内存管理

multiple programming paradigm	多程序设计范式
standard library	标准库
multi-paradigm programming language	多范式编程语言
structured programming	结构化程序设计
aspect-oriented programming	面向方面编程
design by contract	契约式设计
a combination of ...	……的组合
cycle-detecting garbage collector	循环检测垃圾收集器
memory management	内存管理
dynamic name resolution	动态名称解析
variable name	变量名
curly bracket	花括号；大括号
off-side rule	缩排规则
control flow	控制流
assignment statement	赋值语句
local variable	局部变量
as long as	只要；如果
object-oriented programming	面向对象程序设计
code block	代码块
generator function	生成器函数
infix operator	插入算符，中缀运算符
list comprehension	列表推导式
generator expression	生成器推导式
conditional expression	条件表达式
string format	字符串格式
quote mark	引号
escape character	转义字符
a series of	一系列；一连串
raw string	原始字符串
Internet-facing application	面向因特网的应用
pseudorandom number	伪随机数
regular expression	正则表达式
unit testing	单元测试
test suite	测试套件
image processing	图像加工，图像处理

command line interpreterfor	命令行解释程序
session state retention	会话状态保留

Abbreviations

NOP (NoOperation)	无操作
MIME(Multipurpose Internet Mail Extensions)	多用途互联网邮件扩展
HTTP (HyperText Transfer Protocol)	超文本传输协议
REPL (Read–Eval–Print Loop)	读取—求值—输出循环

Text B 参考译文

Python 编程语言

Python 具有动态类型系统和自动内存管理功能。它支持多种编程范例，包括面向对象、命令式、函数和过程，并且具有大型且全面的标准库。

1. 特点和哲学

Python 是一种多范式编程语言。完全支持面向对象编程和结构化编程，其许多功能支持函数式编程和面向方面编程。通过扩展支持许多其他范例，包括契约式设计和逻辑编程。

Python 使用动态类型，并组合引用计数和循环检测垃圾收集器来进行内存管理。它还具有动态名称解析（后期绑定）的特点，它在程序执行期间绑定方法和变量名称。

2. 句法和语义

Python 是一种公认的易于阅读的语言。它的格式看起来整洁，并且它通常在其他语言使用标点符号的地方使用英语关键字。与许多其他语言不同，它不使用大括号来分隔块，而且语句后面的分号是可选的。它比 C 或 Pascal 具有更少的语法异常和特殊情况。

2.1 缩进

Python 使用空格缩进，而不是使用大括号或关键字来界定块。某些语句之后会出现缩进增加的情况；缩进减少表示当前块的结束。因此，程序的视觉结构准确地表示了程序的语义结构。此功能有时也被称为缩排规则。

2.2 语句和控制流程

Python 的语句包括：

- 赋值语句（标记"="，等号）。
- if 语句，有条件地执行代码块，也可以与 else 和 elif（else-if 的缩写）一起使用。
- for 语句，迭代可迭代对象，将每个元素捕获到局部变量以供附加块使用。
- while 语句，只要条件为真，就执行一段代码。
- raise 语句，用于引发指定的异常或重新引发捕获的异常。
- class 语句，它执行代码块并将其本地名称空间附加到类中，以用于面向对象编程。
- def 语句，定义函数或方法。
- pass 语句，用作 NOP。在语法上需要创建一个空代码块。
- assert 语句，在调试期间用于检查需应用的条件。
- yield 语句，它从生成器函数返回一个值。从 Python 2.5 开始，yield 也是一个运算符。
- import 语句，用于导入在当前程序中可用的函数或变量模块。有三种使用导入的方法：import <module name> [as <alias>]或 from <module name> import *或 from <module name> import <definition 1> [as <alias 1>]，<definition 2> [as <alias 2>]，....
- print 语句在 Python 3 中已更改为 print（ ）函数。

2.3 表达式

有些 Python 的表达式与 C 和 Java 等语言类似，而有些则不同。

（1）加法、减法和乘法是相同的，但除法的行为不同。Python 有两种除法类型。Python 还添加了**运算符用于取幂。

（2）从 Python 3.5 开始，引入了新的@中缀运算符。它旨在把诸如 NumPy 的库用于矩阵乘法。

（3）在 Python 中，==按值进行比较，而 Java 中按值比较数字并按照引用比较对象。Python 的 is 运算符可用于比较对象标识。在 Python 中，可以链式比较，例如 *a*<= *b* <= *c*。

（4）Python 在布尔运算中使用单词 and、or、not，而不使用符号&&、||、!，这与 Java 和 C 不同。

（5）Python 有一种称为列表推导的表达式。Python 2.4 将列表推导扩展为更通用的表达式，称为生成器表达式。

（6）Python 中的条件表达式写为 *x* if *c* else *y* （不同于许多其他语言通常使用的 *c*? *x*：*y* 形式）。

（7）Python 区分列表和元组。列表写为[1,2,3]，是可变的，不能用作字典的键（字典键必须在 Python 中不可变）。元组写为（1,2,3），是不可变的，因此可以用作字典的键，前提是元组的所有元素都是不可变的。

（8）Python 有一个"字符串格式"运算符%。该功能类似于 C 中的 printf 格式字符串，例如"spam =%s eggs =%d"%（"blah"，2）。

（9）Python 包含以下各种字符串文字。

- 由单引号或双引号分隔的字符串。这两种字符串都使用反斜杠（\）作为转义字符。
- 三引号字符串，以一系列三个单引号或双引号开头和结尾。
- 原始字符串，用字符串文字前缀 r 表示。不解释转义序列。

3. 库

Python 的大型标准库通常被认为是其最大的优势之一，它提供了适合许多任务的工具。对于面向因特网的应用程序，支持许多标准格式和协议，如 MIME 和 HTTP。它包括用于创建图形用户界面、连接到关系数据库、生成伪随机数、具有任意精度小数的算术、操纵正则表达式和单元测试的模块。

标准库的某些部分包含在规范中，但大多数模块都没有。它们由代码、内部文档和测试套件（如果提供）指定。但是，由于大多数标准库是跨平台的 Python 代码，因此只有少数模块需要更改或重写。

截至 2018 年 3 月，Python Package Index（PyPI），第三方 Python 软件的官方库，包含 130000 多个包，具有广泛的功能，包括：

- 图形用户界面。
- Web 框架。
- 多媒体。
- 数据库。
- 联网。
- 测试框架。
- 自动化。
- 网络抓取。
- 文档建立。
- 系统管理。
- 科学计算。
- 文本处理。
- 图像处理。

4. 开发环境

大多数 Python 实现（包括 CPython）都包含一个读取—求值—输出循环（REPL），允许它们作为命令行解释器运行，用户可以按顺序输入语句并立即接收结果。

其他 shell 增加了更多功能，如自动完成、会话状态保留和语法突出显示。

除标准桌面集成开发环境外，还有基于 Web 浏览器的 IDE：SageMath，用于开发科学和数学相关的 Python 程序；PythonAnywhere，一个基于浏览器的 IDE 和托管环境；以及 Canopy

IDE，一个强调科学计算的商业 Python IDE。

Exercises

[Ex. 1] Answer the following questions according to Text A.

1. What is programming language?

2. How many main categories are programming languages basically classified into? What are they?

3. What are low-level languages?

4. How many categories are low-level languages further classified into? What are they?

5. What is machine language? What does it consist of?

6. What is the third advantage of low-level languages mentioned in the passage?

7. What are high-level languages?

8. How many categories are high-level languages grouped into based on execution model? What can they also be classifies into on the basis of programming paradigm?

9. What is the first disadvantages of high-level language mentioned in the passage?

10. In what aspects does the author talk about the differences between low-level language and high-level language?

[Ex. 2] Fill in the following blanks according to Text B.

1. Python features ________________ and ________________. It supports multiple programming paradigms, including ____________, ____________, ________________ and procedural, and has a large and comprehensive ________________.

2. Python uses ________________, and a combination of ________________ and ________________ for memory management.

3. Python is meant to be ________________. Its formatting is ________________, and it often uses ________________ where other languages use punctuation.

4. Python uses ________________, rather than curly brackets or__________, to delimit blocks. An increase in indentation comes after ________________; a decrease in indentation signifies ________________.

5. The for statement iterates over ____________, capturing each element to a local variable for use by ____________.

6. The import statement is used to ____________ whose functions or ____________ can be used in the current program. There are ____________ ways of using import.

7. Some Python expressions are ____________ languages such as C and Java, while some are not. Python uses the words and, or, not for ____________rather than the symbolic &&, ||, ! used in ____________. Python has a type of expression termed ____________.

8. Python makes a distinction between ____________ and ____________. Lists are written as [1, 2, 3], are__________, and cannot be used as ____________________. Tuples are written as

__________, are immutable and thus can be used as the keys of dictionaries, provided __________________________.

9. Some parts of the standard library are covered by ____________, but most modules are not. They are specified by ____________, ____________, and ____________

10. Most Python implementations (including CPython) include ______________, permitting them to function as ________________ for which the user enters statements sequentially and ________________.

[Ex. 3] Translate the following terms or phrases from English into Chinese and vice versa.

1. embedded system	1. ________________
2. instruction set	2. ________________
3. memory leaks	3. ________________
4. structured programming language	4. ________________
5. control flow	5. ________________
6. *n.*抽象；抽象概念	6. ________________
7. *n.*编译	7. ________________
8. *n.*执行，实行	8. ________________
9. *n.*便携性	9. ________________
10. *n.*解释，说明	10. ________________

[Ex. 4] Translate the following sentences into Chinese.

OOP

Object-oriented programming (OOP) refers to a type of computer programming (software design) in which programmers define the data type of a data structure, and also the types of operations (functions) that can be applied to the data structure.

In this way, the data structure becomes an object that includes bothdata and functions. In addition, programmers can create relationships between one object and another. For example, objects can inherit characteristics from other objects.

1. The Basic OOP Concepts

If you are new to object-oriented programming languages, you will need to know a few basics before you can get started with code. The following definitions will help you better understand object-oriented programming:

- Abstraction: The process of picking out (abstracting) common features of objects and procedures.
- Class: A category of objects. The class defines all the common properties of the different objects that belong to it.

- Encapsulation: The process of combining elements to create a new entity. A procedure is a type of encapsulation because it combines a series of computer instructions.
- Information hiding: The process of hiding details of an object or function. Information hiding is a powerful programming technique because it reduces complexity.
- Inheritance: A feature that represents the "is a" relationship between different classes.
- Interface: The languages and codes that the applications use to communicate with each other and with the hardware.
- Messaging: Message passing is a form of communication used in parallel programming and object-oriented programming.
- Object: A self-contained entity that consists of both data and procedures to manipulate the data.
- Polymorphism: A programming language's ability to process objects differently depending on their data type or class.
- Procedure: A section of a program that performs a specific task.

2. Advantages of Object-Oriented Programming

One of the principal advantages of object-oriented programming techniques over procedural programming techniques is that they enable programmers to create modules that do not need to be changed when a new type of object is added. A programmer can simply create a new object that inherits many of its features from existing objects. This makes object-oriented programs easier to modify.

[Ex. 5] Fill in the blanks with the words given below.

platforms	portable	syntax	compilation	modules
source	loading	solutions	threading	encapsulation

Python Features

1. Interpreted Language

Python is an interpreted language i.e. interpreter executes the code line by line at a time. When you use an interpreted language like Python, there is no separate ___1___ and execution steps. You just run the program from the ___2___ code. This makes debugging easy and thus suitable for beginners. Internally, Python converts the source code into an intermediate form called bytecodes and then translates this into the native language of your specific computer and then runs it. You just run your programs and you never have to worry about linking and ___3___ with libraries, etc.

2. Cross-platform Language

Python can run equally on different ___4___ such as Windows, Linux, Unix, Macintosh etc. A Python program written on a Macintosh computer will run on a Linux system and vice versa. Thus, Python is a ___5___ language.

3. Object-Oriented Language

Python supports object-oriented features. Compared with other programming languages, Python's class mechanism adds classes with a minimum of new ___6___ and semantics. Python classes provide all the standard features of object oriented programming language except strong ___7___, which is only one of many features associated with the term "object-oriented".

4. Extensive Libraries

The Python standard library is huge indeed. Python library contains built-in ___8___ (written in C) that provide access to system functionality such as file I/O that would otherwise be inaccessible to Python programmers, as well as modules written in Python that provide standardized ___9___ for many problems that occur in everyday programming. It can help you do various things involving regular expressions, documentation generation, unit testing, ___10___, databases etc.

Online Resources

二维码	内　容
	计算机专业常用语法（7）：倒装句
	在线阅读（1）：Software Project Management （软件项目管理）
	在线阅读（2）：Programming Languages You Should Learn in 2020 （2020 年你应该学习的编程语言）

Unit 8

Computer Network

Text A

Computer Network Basic

1. What Is a Computer Network

Computer network is a group of computers connected with each other through wires, optical fiber or optical links so that various devices can interact with each other through a network. The aim of computer network is to share resources among various devices.

2. Components of Computer Network

2.1 NIC

NIC is a device that helps the computer to communicate with another device. The network interface card contains the hardware addresses. The data link layer protocol use this address to identify the system on the network so that it transfers the data to the correct destination.

There are two types of NIC: wireless NIC and wired NIC.

- Wireless NIC: All the modern laptops use the wireless NIC. In wireless NIC, a connection is made using the antenna that employs the radio wave technology.
- Wired NIC: Cables use the wired NIC to transfer the data over the medium.

2.2 Hub

Hub is a central device that splits the network connection into multiple devices. When a computer requests for information from another computer, it sends the request to the hub. Hub distributes this request to all the interconnected computers.

2.3 Switch

Switch is a networking device that groups all the devices over the network to transfer the data to another device. A switch is better than a hub as it does not broadcast the message over the network, i.e., it sends the message to the device for which it belongs to. Therefore, we can say that switch sends the message directly from source to the destination.

2.4 Cable

Cable is a transmission media that transmits the communication signals. There are three types of cables:

- Twisted pair cable: It is a high-speed cable that transmits the data over 1Gbps or more.
- Coaxial cable: It resembles like a TV installation cable. It is more expensive than twisted pair cable, but it provides the high data transmission speed.
- Fibre optic cable: It is a high-speed cable that transmits the data using light beams. It provides higher data transmission speed as compared to other cables. It is more expensive as compared to other cables.

2.5 Router

Router is a device that connects the LAN to the internet. The router is mainly used to connect the distinct networks or connect the internet to multiple computers.

3. Uses of Computer Network

- Resource sharing: It is the sharing of resources such as programs, printers, and data among the users on the network without the requirement of the physical location of the resource and user.
- Server-Client model: Computer networking is used in the server-client model. A server is a central computer used to store the information and maintained by the system administrator. Clients are the machines used to access the information stored in the server remotely.
- Communication medium: Computer network behaves as a communication medium among the users. For example, a company contains more than one computer and has an email system which the employees use for daily communication.
- E-commerce: Computer network is also important in businesses. People can do business over the internet. For example, amazon.com is doing their business over the internet.

4. Computer Network Architecture

Computer network architecture is defined as the physical and logical design of the software, hardware, protocols, and media of the transmission of data. There are two types of network architectures used: peer-to-peer network and client/server network.

4.1 Peer-to-Peer Network

Peer-to-peer network is a network in which all the computers are linked together with equal privilege and responsibilities for processing the data. It is useful for small environments, usually up to 10 computers. It has no dedicated server. Special permissions are assigned to each computer for sharing the resources, but this can lead to a problem if the computer with the resource is down.

Advantages of peer-to-peer network:

- It is less costly as it does not contain any dedicated server.
- It is easy to set up and maintain as each computer manages itself.

Disadvantages of peer-to-peer network:

- In the case of peer-to-peer network, it does not contain the centralized system. Therefore, it cannot back up the data as the data is different in different locations.
- It has a security issue as the device is managed itself.

4.2 Client/Server Network

Client/server network is a network model designed for the end users called clients to access the resources from a central computer known as server.

The central controller is known as a server while all other computers in the network are called clients. A server performs all the major operations such as security and network management. It is responsible for managing all the resources such as files, directories, printer, etc. All the clients communicate with each other through a server. For example, if client1 wants to send some data to client 2, then it first sends the request to the server for the permission. The server sends the response to the client 1 to initiate its communication with the client 2.

Advantages of client/server network:

- It contains the centralized system. Therefore, we can back up the data easily.
- It has a dedicated server that improves the overall performance of the whole system.
- Security is better in client/server network as a single server administers the shared resources.
- It also increases the speed of the sharing resources.

Disadvantages of client/server network:

- It is expensive as it requires the server with large memory.
- A server has a network operating system (NOS) to provide the resources to the clients, but the cost of NOS is very high.
- It requires a dedicated network administrator to manage all the resources.

5. Features of Computer Network

5.1 Communication Speed

Network allows us to communicate over the network in a fast and efficient manner. For example, we can do video conferencing, email messaging, etc. over the internet. Therefore, the computer network is a great way to share our knowledge and ideas.

5.2 File Sharing

File sharing is one of the major advantages of the computer network. Computer network allows us to share the files with each other.

5.3 Easy Back up

Since the files are stored in the main server which is centrally located, it is easy to take the back up from the main server.

5.4 Software and Hardware Sharing

We can install the applications on the main server, so the user can access the applications centrally. We do not need to install the software on every machine. Similarly, hardware can also be shared.

5.5 Security

Network offers the security by ensuring that the user has the right to access the certain files and applications.

5.6 Scalability

Scalability means that we can add new components on the network. Network must be scalable so that we can extend the network by adding new devices. But, it decreases the speed of the connection and data of the transmission speed also decreases. This increases the chances of error occurring. This problem can be overcome by using the routing or switching devices.

5.7 Reliability

Computer network can use the alternative source for the data communication in case of any hardware failure.

6. Computer Network Types

6.1 LAN (Local Area Network)

Local area network is a group of computers connected to each other in a small area such as building, office etc. It is used for connecting two or more personal computers through a communication medium such as twisted pair, coaxial cable, etc. It is less costly as it is built with inexpensive hardware such as hubs, network adapters, and Ethernet cables. The data is transferred at an extremely fast rate in local area network. Local area network provides higher security.

6.2 MAN (Metropolitan Area Network)

A metropolitan area network is a network that covers a larger geographic area by interconnecting different LANs to form a larger network. Government agencies use MAN to connect to the citizens and private industries. In MAN, various LANs are connected to each other through a telephone exchange line. It has a higher range than LAN.

Uses of metropolitan area network:

- MAN is used in communication between the banks in a city.
- It can be used in an airline reservation.
- It can be used in a college within a city.

6.3 WAN (Wide Area Network)

A wide area network is a network that extends over a large geographical area such as states or countries. It is quite bigger network than the LAN. It is not limited to a single location, but it spans over a large geographical area through a telephone line, fibre optic cable or satellite links. The internet is one of the biggest WAN in the world. A wide area network is widely used in the field of business, government, and education.

Examples of wide area network:

- Mobile broadband: A 4G network is widely used across a region or country.
- Last mile: A telecom company provides the internet services to the customers in hundreds of cities by connecting their home with fiber.
- Private network: A bank provides a private network that connects lots of offices. This network is made by using the telephone leased line provided by the telecom company.

Advantages of wide area network:

- Geographical area: A wide area network provides a large geographical area. Suppose if the branch of our office is in a different city then we can connect with them through WAN. The internet provides a leased line through which we can connect with another branch.
- Centralized data: In case of WAN network, data is centralized. Therefore, we do not need to buy the emails, files or back up servers.
- Get updated files: Software companies work on the live server. Therefore, the programmers get the updated files within seconds.
- Exchange messages: In a WAN network, messages are transmitted fast. The web application like Facebook, Whatsapp, Skype allows you to communicate with friends.
- Sharing of software and resources: In WAN network, we can share the software and other resources like a hard drive, RAM.
- Global business: We can do the business over the internet globally.

Disadvantages of wide area network:

- Security issue: A WAN network has more security issues as compared to LAN and MAN network, as all the technologies are combined together and that creates the security problem.
- Need of firewall and antivirus software: The data is transferred on the internet which can be changed or hacked by the hackers, so firewall needs to be used. Some people can inject the virus in our system so antivirus is needed .
- High setup cost: An installation cost of the WAN network is high as it involves the purchasing of routers, switches.
- Troubleshooting problems: It covers a large area so fixing the problem is difficult.

6.4 Internetwork

An internetwork is defined as two or more LANs or WANs or computer network segments which are connected using devices and they are configured by a local addressing scheme. An interconnection between public, private, commercial, industrial, or government computer networks can be defined as internetworking. An internetworking uses the internet protocol. The reference model used for internetworking is open system interconnection (OSI).

There are two types of internetwork. They are extranet and intranet.

An extranet is a communication network based on the internet protocol. It is used for information sharing. The access to the extranet is restricted to only those users who have login credentials. An extranet is the lowest level of internetworking. It can be categorized as MAN, WAN or other computer networks. An extranet cannot have a single LAN. It must have at least one connection to the external network.

An intranet is a private network based on the internet protocol. An intranet belongs to an organization which is only accessible by the organization's employee or members. The main aim of the intranet is to share the information and resources among the organization employees. An intranet provides the facility to work in groups and for teleconferences.

New Words

network	[ˈnetwɜːk]	*n.*网络
		*v.*将……连接成网络
wire	[ˈwaɪə]	*n.*电线
link	[lɪŋk]	*n.*链接
address	[əˈdres]	*n.*地址
protocol	[ˈprəʊtəkɒl]	*n.*协议
transfer	[trænsˈfɜː]	*vt.*传输
antenna	[ænˈtenə]	*n.*天线
radio	[ˈreɪdɪəʊ]	*n.*无线电
medium	[ˈmiːdɪəm]	*n.*介质，媒介物
		*adj.*中等的，中级的
hub	[hʌb]	*n.*集线器
split	[splɪt]	*vt.*分开；分担
		*n.*划分
request	[rɪˈkwest]	*n.&vt.*请求
distribute	[dɪˈstrɪbjuːt]	*vt.*分配，散发，分发
switch	[swɪtʃ]	*n.*交换机
broadcast	[ˈbrɔːdkɑːst]	*vt.*广播
transmission	[trænsˈmɪʃn]	*n.*播送；传送
signal	[ˈsɪgnəl]	*n.*信号
		*vt.*向……发信号

		*vi.*发信号
resemble	[rɪˈzembl]	*vt.*与……相像，类似于
router	[ˈru:tə]	*n.*路由器
administrator	[ədˈmɪnɪstreɪtə]	*n.*管理员，管理者
E-commerce	[i:'kɒmɜ:s]	*n.*电子商务
logical	[ˈlɒdʒɪkl]	*adj.*逻辑的
privilege	[ˈprɪvəlɪdʒ]	*n.*特权
permission	[pəˈmɪʃn]	*n.*允许；批准
initiate	[ɪˈnɪʃieɪt]	*vt.*开始，发起
dedicated	[ˈdedɪkeɪtɪd]	*adj.*专用的
scalable	['skeɪləbl]	*adj.*可扩展的
decrease	[dɪˈkri:s]	*v.*降低；减少，减小
overcome	[ˌəʊvəˈkʌm]	*v.*战胜，克服
alternative	[ɔ:lˈtɜ:nətɪv]	*adj.*替代的；备选的
		*n.*可供选择的事物
adapter	[ə'dæptə]	*n.*适配器
geographic	[ˌdʒi:ə'græfɪk]	*adj.*地理的
span	[spæn]	*n.*跨度
satellite	[ˈsætəlaɪt]	*n.*卫星；人造卫星
broadband	[ˈbrɔ:dbænd]	*n.*宽带
antivirus	[ˈæntɪvaɪrəs]	*adj.*抗病毒的，防病毒的
inject	[ɪnˈdʒekt]	*vt.*引入；（给……）添加
fix	[fɪks]	*vt.*维修，修理；准备
internetwork	[ɪntɜ:r'netwɜ:k]	*n.*互联网络
segment	[ˈsegmənt]	*n.*部分，段落
		*v.*分割，划分
extranet	['ekstrənet]	*n.*外联网
intranet	[ˈɪntrənet]	*n.*内联网
teleconference	[ˈtelɪkɒnfərəns]	*n.*远程电信会议

Phrases

optical fiber	光纤，光缆
data link layer	数据链路层
radio wave	无线电波
networking device	网络设备

twisted pair	双绞线
coaxial cable	同轴电缆
data transmission speed	数据传输速度
fibre optic cable	光缆
light beam	光束
physical location	物理位置
peer-to-peer network	对等网
set up	建立；准备；安排
end user	终端用户
video conferencing	视频会议
file sharing	文件共享，文件分享
communication medium	通信介质
last mile	最后路程，最后一英里
antivirus software	防病毒软件
be restricted to	仅限于，局限于
be categorized as	归类为

Abbreviations

NIC (Network Interface Controller)	网络接口卡，网络适配器
bps (bits per second)	每秒位
NOS (Network Operating System)	网络操作系统
LAN (Local Area Network)	局域网
MAN (Metropolitan Area Network)	城域网
WAN (Wide Area Network)	广域网
OSI (Open System Interconnection)	开放系统互联模型

Text A 参考译文

计算机网络基础

1. 什么是计算机网络

计算机网络是一组通过电线、光纤或光链路相互连接的计算机，以便各种设备可以通过网络交互。计算机网络的目的是在各种设备之间共享资源。

2. 计算机网络的组成部分

2.1 网络接口卡

网络接口卡是帮助计算机与另一台设备通信的设备。网络接口卡包含硬件地址。数据链路层协议使用该地址来标识网络上的系统，以便将数据传输到正确的目的地。

网络接口卡有两种类型：无线网络接口卡和有线网络接口卡。

- 无线网络接口卡：所有现代笔记本电脑都使用无线网络接口卡。在无线网络接口卡中，使用采用无线电波技术的天线进行连接。
- 有线网络接口卡：电缆使用有线网络接口卡在介质上传输数据。

2.2 集线器

集线器是将网络连接拆分为多个设备的中央设备。当计算机向另一个计算机请求信息时，它将请求发送到集线器。集线器将此请求分发给所有互连的计算机。

2.3 交换机

交换机是一种联网设备，它将网络上的所有设备分组以将数据传输到另一台设备。交换机比集线器更好，因为它不通过网络广播消息，即它将消息发送到它所属的设备。因此，可以说交换机将消息直接从源发送到目的地。

2.4 电缆

电缆是传输通信信号的传输介质。电缆分为三种。

- 双绞线：这是一条高速电缆，用于以 1Gbps 或更高的速率传输数据。
- 同轴电缆：同轴电缆类似于电视安装的电缆。它比双绞线电缆贵，但是它提供了高的数据传输速度。
- 光纤电缆：光纤电缆是使用光束传输数据的高速电缆。与其他电缆相比，它提供了更高的数据传输速度。它比其他电缆更贵。

2.5 路由器

路由器是将局域网连接到互联网的设备。路由器主要用于连接不同的网络或将互联网连接到多台计算机。

3. 计算机网络的用途

- 资源共享：资源共享是指网络上的用户之间共享程序、打印机和数据之类的资源，而不用管资源和用户的物理位置。
- 服务器——客户端模型：在服务器——客户端模型中使用了的计算机网络。服务器是用于存储信息并由系统管理员维护的中央计算机。客户端是用于远程访问服务器中存储的信息的计算机。

- 通信介质：计算机网络充当用户之间的通信介质。例如，一家公司拥有一台以上的计算机，并且有一个电子邮件系统，员工可以使用该电子邮件系统进行日常通信。
- 电子商务：计算机网络在企业中也很重要。人们可以通过互联网开展业务。例如，amazon.com 正在通过互联网开展业务。

4. 计算机网络体系结构

计算机网络体系结构被定义为软件、硬件、协议和数据传输介质的物理和逻辑设计。有两类网络体系结构：对等网络和客户端/服务器网络。

4.1 对等网络

对等网络是一种网络，其中所有计算机都以相同的优先权和责任连接在一起，以处理数据。它通常用于小型环境，最多有 10 台计算机。它没有专用服务器。每台计算机都分配了特殊权限以共享资源，但是如果资源所在的计算机关闭，这可能会引发问题。

对等网络的优点：

- 它不包含任何专用服务器，因此成本较低。
- 由于每台计算机都可以自我管理，因此易于设置和维护。

对等网络的缺点：

- 对于对等网络，它不包含集中式系统。因此，由于不同位置的数据不同，无法备份数据。
- 由于设备自我管理，因此存在安全问题。

4.2 客户端/服务器网络

客户端/服务器网络是为终端用户（称为客户端）设计的网络模型，可从称为服务器的中央计算机访问资源。

中央控制器称为服务器，而网络中的所有其他计算机称为客户端。服务器执行所有主要操作，例如安全性和网络管理。它负责管理所有资源，例如文件、目录、打印机等。所有客户端都通过服务器相互通信。例如，如果 client 1 要向 client 2 发送一些数据，则它首先将请求发送到服务器以获取许可。服务器将响应发送到 client 1，以启动与 client 2 的通信。

客户端/服务器网络的优点：

- 它包含集中式系统。因此，可以轻松备份数据。
- 它具有专用的服务器，可以提高整个系统的整体性能。
- 它的安全性更高，因为由一台服务器管理共享资源。
- 它还可以提高共享资源的速度。

客户端/服务器网络的缺点：

- 客户端/服务器网络价格昂贵，因为它要求服务器具有大内存。
- 服务器具有网络操作系统（NOS）为客户端提供资源，但是 NOS 的成本很高。
- 它需要专门的网络管理员来管理所有资源。

5. 计算机网络的特点

5.1 通信速度

网络使我们能够快速有效地通过网络进行通信。例如，我们可以通过互联网进行视频会议，用电子邮件传递消息等。因此，计算机网络是共享知识和思想的好方法。

5.2 文件共享

文件共享是计算机网络的主要优势之一。计算机网络使我们可以相互共享文件。

5.3 备份容易

由于文件存储在位于中央的主服务器上。因此，很容易从主服务器上进行备份。

5.4 软件和硬件共享

可以在主服务器上安装应用程序，这样，用户可以集中访问应用程序。不需要在每台机器上都安装软件。同样，硬件也可以共享。

5.5 安全性

网络通过确保用户有权访问某些文件和应用程序来提供安全性。

5.6 可扩展性

可扩展性意味着可以在网络上添加新组件。网络必须具有可扩展性，以便可以通过添加新设备来扩展网络。但是，它降低了连接速度，并且数据的传输速度也降低了，这增加了发生错误的机会。通过使用路由或交换设备可以解决此问题。

5.7 可靠性

万一发生任何硬件故障，计算机网络可以使用备用源进行数据通信。

6. 计算机网络类型

6.1 LAN（局域网）

局域网是指在建筑物、办公室等较小区域内相互连接的一组计算机。它用于通过通信介质（如双绞线、同轴电缆等）连接两个或更多个人计算机。它由廉价的硬件（例如集线器、网络适配器和以太网电缆）构建，因此成本较低。数据在局域网中的传输速度非常快。局域网提供更高的安全性。

6.2 城域网

城域网是指通过把不同的局域网连接起来组成的覆盖区域更大的网络。政府机构使用城域网连接到公民和私营企业。在城域网中，各种局域网通过电话交换线路相互连接。它具有比局域网更大的范围。

城域网的用途：

- 城域网用于城市中银行之间的通信。
- 可用于航空公司订票系统。
- 可用于城市中的大学。

6.3 广域网

广域网是延伸到很大的地理区域（例如州或国家/地区）的网络。广域网是比局域网更大的网络。广域网不限于单个位置，而是通过电话线、光纤电缆或卫星链路跨越很大的地理区域。互联网是世界上最大的广域网之一。广域网广泛用于商业、政府和教育领域。

广域网的例子有：

- 移动宽带：4G 网络在一个地区或国家/地区广泛使用。
- 最后一英里：一家电信公司用光纤将房屋连接起来，为数百个城市的客户提供互联网服务。
- 专用网络：银行提供连接多个办公室的专用网络。该网络使用电信公司提供的电话租用线路建立。

广域网的优点：

- 地理区域：广域网提供很大的地理区域。假设我们办公室的分支机构位于其他城市，则可以通过广域网与之连接。互联网提供了一条专线，可以通过它与另一个分支机构连接。
- 集中数据：在广域网中，数据是集中的。因此，不需要购买电子邮件、文件或备份服务器。
- 获取更新的文件：软件公司在实时服务器上工作。因此，程序员可以在几秒钟内获得更新的文件。
- 交换消息：在广域网中，消息可以快速传输。Facebook、Whatsapp、Skype 等网络应用程序使你可以与朋友交流。
- 软件和资源共享：在广域网中，我们可以共享软件和其他资源，例如硬盘驱动器、RAM。
- 全球业务：我们可以在全球范围内通过互联网开展业务。

广域网的缺点：

- 安全问题：与局域网和城域网相比，广域网存在更多的安全问题，因为所有技术组合在一起就产生了安全问题。

- 需要防火墙和防病毒软件：数据在互联网上传输，可以被黑客更改或入侵，因此需要使用防火墙。某些人可以将病毒注入我们的系统中，因此需要防病毒。
- 高昂的安装成本：由于需要购买路由器、交换机，因此广域网的安装成本很高。
- 故障排除问题：它覆盖的区域很大，因此解决问题很难。

6.4 互联网络

把两个或多个局域网或广域网或计算机网段用设备连接起来就被定义为互联网络，通过本地编址方案进行配置。公共、私有、商业、工业或政府计算机网络之间的互联也可以定义为互联网络。互联网络使用网际协议。用于互联网络的参考模型是开放系统互联（OSI）。

互联网络有两类：外联网和内联网。

外联网是基于网际协议的通信网络。它用于信息共享。只有有登录凭据的用户才能访问外联网。外联网是最低级别的互联网络。它可以归类为城域网、广域网或其他计算机网络。一个外联网不能只有一个局域网，它必须与外部网络建立至少一个连接。

内联网是基于网际协议的专用网络。内联网属于某个组织，只让组织的员工或成员访问。内联网的主要目的是在组织的员工之间共享信息和资源。内联网提供了用于分组工作和远程电信会议的功能。

Text B

扫码听课文

Computer Network Topology

1. Bus Topology

Bus topology is designed in such a way that all the stations are connected through a single cable known as a backbone cable. Each node is either connected to the backbone cable by drop cable or directly connected to the backbone cable. When a node wants to send a message over the network, it puts a message over the network. All the stations available in the network will receive the message whether it has been addressed or not. The bus topology is mainly used in 802.3 (Ethernet) and 802.4 standard networks. The configuration of a bus topology is quite simpler as compared to other topologies. The backbone cable is considered as a "single lane" through which the message is broadcast to all the stations. The most common access method of the bus topologies is CSMA (Carrier Sense Multiple Access).

Advantages of bus topology:

- Low-cost cable: In bus topology, nodes are directly connected to the cable without passing through a hub. Therefore, the initial cost of installation is low.
- Moderate data speeds: Coaxial or twisted pair cables are mainly used in bus-based networks that support upto 10 Mbps.
- Familiar technology: Bus topology is a familiar technology as the installation and

troubleshooting techniques are well known, and hardware components are easily available.
Limited failure: A failure in one node will not have any effect on other nodes.

- Disadvantages of bus topology:
- Extensive cabling: A bus topology is quite simple, but it still requires a lot of cabling.
- Difficult troubleshooting: It requires specialized test equipment to determine the cable faults. If any fault occurs in the cable, then it would disrupt the communication for all the nodes.
- Signal interference: If two nodes send the messages simultaneously, then the signals of both the nodes collide with each other.
- Reconfiguration difficult: Adding new devices to the network would slow down the network.
- Attenuation: Attenuation is a loss of signal which leads to communication issues. Repeaters are used to regenerate the signal.

2. Ring Topology

Ring topology is like a bus topology, but with connected ends. The node that receives the message from the previous computer will retransmit it to the next node. The data flows in one direction, i.e., it is unidirectional. The data flows in a single loop continuously known as an endless loop. It has no terminated ends, i.e., each node is connected to other node and having no termination point. The data in a ring topology flow in a clockwise direction.

The most common access method of the ring topology is token passing. It is a network access method in which token is passed from one node to another node. Token is a frame that circulates around the network.

Advantages of ring topology:

- Network management: Faulty devices can be removed from the network without bringing the network down.
- Product availability: Many hardware and software tools for network operation and monitoring are available.
- Cost: Twisted pair cabling is inexpensive and easily available. Therefore, the installation cost is very low.
- Reliable: It is a more reliable network because the communication system is not dependent on the single host computer.

Disadvantages of ring topology:

- Difficult troubleshooting: It requires specialized test equipment to determine the cable faults. If any fault occurs in the cable, then it would disrupt the communication for all the nodes.
- Failure: The breakdown in one station leads to the failure of the overall network.
- Difficult reconfiguration : Adding new devices to the network would slow down the network.

- Delay: Communication delay is directly proportional to the number of nodes. Adding new devices increases the communication delay.

3. Star Topology

Star topology is an arrangement of the network in which every node is connected to the central hub, switch or a central computer. The central computer is known as a server, and the peripheral devices attached to the server are known as clients. Coaxial cable or RJ-45 cables are used to connect the computers. Hubs or switches are mainly used as connection devices in a physical star topology. Star topology is the most popular topology in network implementation.

Advantages of star topology:

- Efficient troubleshooting: Troubleshooting is quite efficient in a star topology as compared to bus topology. In a bus topology, the manager has to inspect kilometers of cable to troubleshoot a problem. In a star topology, all the stations are connected to the centralized network. Therefore, the network administrator only goes to the single station to troubleshoot the problem.
- Network control: Complex network control features can be easily implemented in the star topology. Any changes made in the star topology are automatically accommodated.
- Limited failure: As each station is connected to the central hub with its own cable, failure in one cable will not affect the entire network.
- Familiar technology: Star topology is a familiar technology .
- Easily expandable: It is easily expandable as new stations can be added to the open ports on the hub.
- Cost effective: Star topology networks are cost-effective as it uses inexpensive cable.
- High data speeds: It supports a bandwidth of approx 100Mbps. Ethernet 100Base-T is one of the most popular star topology networks.

Disadvantages of star topology:

A central point of failure: If the central hub or switch goes down, then all the connected nodes will not be able to communicate with each other.

4. Tree Topology

A tree topology is a type of structure in which all the computers are connected with each other in hierarchical fashion. The top-most node in tree topology is known as a root node, and all other nodes are the descendants of the root node. There is only one path exists between two nodes for the data transmission. Thus, it forms a parent-child hierarchy.

Advantages of tree topology:

- Support for broadband transmission: Tree topology is mainly used to provide broadband transmission, i.e., signals are sent over long distances without being attenuated.
- Easily expandable: We can add the new device to the existing network. Therefore, we can say that tree topology is easily expandable.

- Easily manageable: In tree topology, the whole network is divided into segments known as star networks which can be easily managed and maintained.
- Error detection: Error detection and error correction are very easy in a tree topology.
- Limited failure: The breakdown in one station does not affect the entire network.
- Point-to-point wiring: It has point-to-point wiring for individual segments.

Disadvantages of tree topology:

- Difficult troubleshooting: If any fault occurs in the node, then it becomes difficult to troubleshoot the problem.
- High cost: Devices required for broadband transmission are very costly.
- Failure: A tree topology mainly relies on main bus cable and failure in main bus cable will damage the overall network.
- Difficult reconfiguration: If new devices are added, it becomes difficult to reconfigure.

5. Mesh Topology

Mesh topology is an arrangement of the network in which computers are interconnected with each other through various redundant connections. There are multiple paths from one computer to another computer. It does not contain the switch, hub or any central computer which acts as a central point of communication. The Internet is an example of the mesh topology.

Mesh topology is mainly used for WAN implementations where communication failures are a critical concern. It is mainly used for wireless networks. It can be formed by using the formula: Number of cables = $(n\times(n-1))/2$; Where n is the number of nodes that represents the network.

Mesh topology is divided into two categories:

- Full mesh topology: In a full mesh topology, each computer is connected to all the computers available in the network.
- Partial mesh topology: In a partial mesh topology, not all but certain computers are connected to those computers with which they communicate frequently.

Advantages of mesh topology:

- Reliable: The mesh topology networks are very reliable. No link breakdown will affect the communication between connected computers.
- Fast Communication: Communication is very fast between the nodes.
- Easier Reconfiguration: Adding new devices would not disrupt the communication between other devices.

Disadvantages of mesh topology:

- Cost: A mesh topology contains a large number of connected devices such as a router and more transmission media than other topologies.
- Management: Mesh topology networks are very large and very difficult to maintain and manage. If the network is not monitored carefully, then the communication link failure goes undetected.
- Efficiency: In this topology, redundant connections are high, which reduces the efficiency of the network.

6. Hybrid Topology

When two or more different topologies are combined together it is known as a hybrid topology. For example, if there exists a ring topology in one branch of ABC bank and bus topology in another branch of ABC bank, connecting these two topologies will result in a hybrid topology.

Advantages of hybrid topology:

- Reliable: If a fault occurs in any part of the network it will not affect the functioning of the rest of the network.
- Scalable: Size of the network can be easily expanded by adding new devices without affecting the functionality of the existing network.
- Flexible: This topology is very flexible as it can be designed according to the requirements of the organization.
- Effective: Hybrid topology is very effective as it can be designed in such a way that the strength of the network is maximized and weakness of the network is minimized.

Disadvantages of hybrid topology:

- Complex design: The major drawback of the hybrid topology is the design of the hybrid network. It is very difficult to design the architecture of the hybrid network.
- Costly hub: The hubs used in the hybrid topology are very expensive as these hubs are different from usual hubs used in other topologies.
- Costly infrastructure: The infrastructure cost is very high as a hybrid network requires a lot of cabling, network devices, etc.

New Words

topology	[tə'pɒlədʒɪ]	*n.*拓扑
bus	[bʌs]	*n.*总线
backbone	[ˈbækbəʊn]	*n.*主干，骨干；脊椎
Ethernet	[ˈi:θənet]	*n.*以太网
station	[ˈsteɪʃn]	*n.*站点 *vt.*配置，安置
moderate	[ˈmɒdərət]	*adj.*中等的；稳健的；适度的
fault	[fɔ:lt]	*n.*故障
disrupt	[dɪsˈrʌpt]	*vt.*使中断；破坏 *adj.*中断的
interference	[ˌɪntəˈfɪərəns]	*n.*干涉，干扰，冲突
simultaneously	[ˌsɪməl'teɪnɪəslɪ]	*adv.*同时地
reconfiguration	['ri:kənfɪgjʊ'reɪʃən]	*n.*重新配置；再配置；再组合
attenuation	[əˌtenjʊ'eɪʃn]	*n.*衰减
repeater	[rɪˈpi:tə]	*n.*中继器，转发器

regenerate	[rɪˈdʒenəreɪt]	*v.*使再生
transmit	[trænsˈmɪt]	*vt.*传输；传送，传递；发射 *vi.*发送信号
continuously	[kən'tɪnjʊəslɪ]	*adv.*连续不断地，接连地
endless	[ˈendlɪs]	*adj.*无尽的，永久的
termination	[ˌtɜːmɪˈneɪʃn]	*n.*结束；终止处
clockwise	[ˈklɒkwaɪz]	*adj.*顺时针方向的
token	[ˈtəʊkən]	*n.*令牌
frame	[freɪm]	*n.*框架；构架
availability	[əˌveɪlə'bɪlɪtɪ]	*n.*可利用性；有效，有益
breakdown	[ˈbreɪkdaʊn]	*n.*崩溃；损坏，故障
proportional	[prəˈpɔːʃənl]	*adj.*比例的，成比例的；相称的
arrangement	[əˈreɪndʒmənt]	*n.*安排；排列
accommodate	[əˈkɒmədeɪt]	*v.*容纳；适应
cost-effective	[kɔst-ɪˈfektɪv]	*adj.*有成本效益的，划算的
expandable	[ɪkˈspændəbl]	*adj.*可扩展的
hierarchical	[ˌhaɪəˈrɑːkɪkl]	*adj.*分层的
descendant	[dɪˈsendənt]	*n.*后代；后裔
manageable	[ˈmænɪdʒəbl]	*adj.*可管理的，易控制的，易处理的
correction	[kəˈrekʃn]	*adj.*改正的，纠正的
redundant	[rɪˈdʌndənt]	*adj.*冗余的，多余的
path	[pɑː θ]	*n.*路径
formula	[ˈfɔːmjʊlə]	*n.*公式，准则
partial	[ˈpɑːʃl]	*adj.*部分的
undetected	[ˌʌndɪˈtektɪd]	*adj.*未被发现的，未探测到的
flexible	[ˈfleksəbl]	*adj.*灵活的；柔韧的
infrastructure	[ˈɪnfrəstrʌktʃə]	*n.*基础设施；基础建设

Phrases

bus topology	总线拓扑
drop cable	分支电缆
single lane	单通道，单行道
a lot of	许多的
collide with	与……相撞，与……相碰
slow down	（使）慢下来；（使）生产缓慢

ring topology	环状拓扑
data flow	数据流
star topology	星状拓扑
tree topology	树状拓扑
root node	根节点
be divided into	被分为
error detection	错误检查
point-to-point wiring	点对点连线
mesh topology	网状拓扑
be formed by	由……组成
transmission media	传输介质
hybrid topology	混合拓扑
the rest of	其余的，剩下的

Abbreviations

CSMA (Carrier Sense Multiple Access)	载波侦听多路访问

Text A 参考译文

计算机网络拓扑

1. 总线拓扑

总线拓扑的设计方式是，所有站点都通过一条称为主干电缆的电缆连接。每个节点通过引入电缆连接到主干电缆或直接连接到主干电缆。当节点想要通过网络发送消息时，它把消息发布到网络。无论是否已寻址，网络中所有可用的站都将收到该消息。总线拓扑主要用于802.3（以太网）和802.4标准网络。与其他拓扑相比，总线拓扑的配置要简单得多。主干电缆被认为是一条"单通道"，通过该通道将消息广播到所有站点。总线拓扑中最常见的访问方法是CSMA（载波侦听多路访问）。

总线拓扑的优点：

- 低成本电缆：在总线拓扑中，节点直接连接到电缆，而无须通过集线器。因此，初始安装成本低。
- 中等数据速率：同轴电缆或双绞线电缆主要用于支持高达10Mbps的基于总线的网络。
- 熟悉的技术：总线拓扑是一种熟悉的技术，因为它的安装和故障排除技术众所周知，并且硬件组件很容易获得。

- 有限的故障：一个节点中的故障不会对其他节点产生任何影响。

总线拓扑的缺点：

- 广泛的布线：总线拓扑结构相当简单，但是它仍然需要大量的布线。
- 难以排除故障：需要专门的测试设备来确定电缆故障。如果电缆发生任何故障，则将中断所有节点的通信。
- 信号干扰：如果两个节点同时发送消息，则两个节点的信号会相互冲突。
- 重新配置困难：将新设备添加到网络会降低网络速度。
- 衰减：衰减是信号丢失，这会导致通信问题。中继器用来重新生成信号。

2. 环形拓扑

环形拓扑就像总线拓扑，但具有连接的两端。从上一台计算机接收消息的节点将把信息转发到下一个节点。数据沿一个方向流动，即单向流动。数据以称为无限循环的连续单向循环不停地流动。它没有终点，即每个节点都连接到另一个节点并且没有端接点。环形拓扑中的数据沿顺时针方向流动。

环形拓扑最常见的访问方法是令牌传递。它是一种网络访问方法，其中令牌从一个节点传递到另一节点。令牌是在网络中流动的框架。

环形拓扑的优点：

- 网络管理：可以从网络中删除故障设备，而不必关闭网络。
- 产品可用性：提供了许多用于网络运行和监管的硬件和软件工具。
- 成本：双绞线电缆价格便宜且易于获得。因此，安装成本非常低。
- 可靠：因为通信系统不依赖于单个主机，所以它是一个更可靠的网络。

环形拓扑的缺点：

- 难以排除故障：需要专门的测试设备来确定电缆故障。如果电缆发生任何故障，则将中断所有节点的通信。
- 失效：一个站点的故障会导致整个网络故障。
- 重新配置困难：将新设备添加到网络会降低网络速度。
- 延迟：通信延迟与节点数成正比。添加新设备会增加通信延迟。

3. 星形拓扑

星形拓扑是网络的一种布置，其中每个节点都连接到中央集线器、交换机或中央计算机。中央计算机称为服务器，连接到服务器的外围设备称为客户端。同轴电缆或 RJ-45 电缆用于连接计算机。集线器或交换机主要用作物理星形拓扑中的连接设备。星形拓扑是网络实施中最流行的拓扑。

星形拓扑的优点：

- 高效的故障排除：与总线拓扑相比，星形拓扑中的故障排除效率很高。在总线拓扑中，管理者必须检查数公里的电缆去排除故障。在星形拓扑中，所有站点都连接到集中式网络。因此，网络管理员只要到单个工作站去解决问题。
- 网络控制：复杂的网络控制功能可以在星形拓扑中轻松实现。星形拓扑中所做的任何更改都将自动适应。
- 有限的故障：由于每个站点都使用自己的电缆连接到中央集线器，因此一根电缆的故障不会影响整个网络。
- 熟悉的技术：星形拓扑是一种熟悉的技术。
- 易于扩展：它可以轻松扩展，因为新工作站可以添加到集线器上的开放端口。
- 经济高效：星形拓扑网络由于使用廉价的电缆而具有成本效益。
- 高数据速度：它支持大约 100Mbps 的带宽。以太网 100Base-T 是最流行的星形拓扑网络之一。

星形拓扑的缺点：

中心点故障：如果中心集线器或交换机出现故障，则所有连接的节点将无法相互通信。

4. 树形拓扑

树形拓扑是一种所有计算机都以分层方式相互连接的结构。树形拓扑中最顶层的节点称为根节点，所有其他节点都是根节点的后代。在两个节点之间只有一条路径可用于数据传输。因此，它形成了父——子层次结构。

树形拓扑的优点：

- 支持宽带传输：树形拓扑主要用于提供宽带传输，即信号在不衰减的情况下长距离发送。
- 易于扩展：可以将新设备添加到现有网络。因此，可以说树形拓扑很容易扩展。
- 易于管理：在树形拓扑中，整个网络被划分为称为星形网络的网段，可以轻松地进行管理和维护。
- 错误检测：在树形拓扑中，错误检测和错误纠正非常容易。
- 有限的故障：一个站点的故障不会影响整个网络。
- 点对点布线：它具有用于各个段的点对点布线。

树形拓扑的缺点：

- 难以排除故障：如果节点中发生任何故障，则很难排除问题。
- 高成本：宽带传输所需的设备非常昂贵。
- 故障：树形拓扑主要依赖于主总线电缆，主总线电缆发生故障会损坏整个网络。
- 难以重新配置：如果添加了新设备，将很难重新配置。

5. 网状拓扑

网状拓扑是网络的一种布置，其中计算机通过各种冗余连接实现互连。从一台计算机到另一台计算机有多种路径。它不包含用作通信中心点的交换机、集线器或任何中央计算机。因特网是网状拓扑的一个例子。

网状拓扑结构主要用于广域网实施，在此，通信故障是人们关注的一个焦点。它主要用于无线网络。其构成符合以下公式：电缆数量=（$n \times (n-1)$）/ 2；其中 n 代表网络的节点数。

网状拓扑分为两类。

- 全网状拓扑：在全网状拓扑中，每台计算机都连接到网络中所有可用的计算机。
- 局部网状拓扑：在局部网状拓扑中，除了某些计算机之外，不是所有计算机都连接到它们经常通信的计算机。

网状拓扑的优点。

- 可靠：网状拓扑网络非常可靠。任何链路故障都不会影响所连接计算机之间的通信。
- 快速通信：节点之间的通信非常快。
- 更容易重新配置：添加新设备不会中断其他设备之间的通信。

网状拓扑的缺点：

- 成本：网状拓扑包含大量的连接设备（例如路由器），并且传输介质比其他拓扑更多。
- 管理：网状拓扑网络非常大，很难维护和管理。如果未仔细监管网络，则无法检测到通信链路故障。
- 效率：在这种拓扑中，冗余连接数很高，从而降低了网络效率。

6. 混合拓扑

将两个或多个不同的拓扑组合在一起时，称为混合拓扑。例如，如果 ABC 银行的一个分行中用环形拓扑，而 ABC 银行的另一分行用总线拓扑，则将这两个拓扑连接起来就是混合拓扑。

混合拓扑的优点：

- 可靠：如果网络的任何部分发生故障，则不会影响网络其余部分的功能。
- 可扩展：可通过添加新设备轻松扩展网络规模，而不会影响现有网络的功能。
- 灵活：此拓扑非常灵活，因为可以根据组织的要求进行设计。
- 有效：混合拓扑非常有效，因为它可以通过设计使网络的性能最优而弱点最小。

混合拓扑的缺点：

- 设计复杂：混合拓扑的主要缺点是混合网络的设计。设计混合网络的架构非常困难。
- 昂贵的集线器：混合拓扑中使用的集线器非常昂贵，因为这些集线器不同于其他拓扑中使用的常规集线器。
- 昂贵的基础架构：基础架构成本很高，因为混合网络需要大量的电缆、网络设备等。

Exercises

[Ex. 1] Answer the following questions according to Text A.

1. What is computer network? What is the aim of computer network?
2. What is NIC? How many types of NIC are there? What are they?
3. What is switch?Why is a switch better than a hub?
4. How many types of cables are there? What are they?
5. What are the uses of computer network?
6. What is peer-to-peer network? What are the advantages of peer-to-peer network?
7. What is a client/server network? What are the disadvantages of client/server network?
8. What are the features of computer network?
9. What is local area network?
10. What are examples of wide area network?

[Ex.2] Answer the following questions according to Text B.

1. How is bus topology designed?
2. What are the advantages of bus topology?
3. What is the most common access method of the ring topology? What are the disadvantages of ring topology?
4. What is star topology? What are hubs or switches mainly used as?
5. What is a tree topology?
6. What are the disadvantages of tree topology?
7. What is mesh topology?
8. What are the advantages of mesh topology?
9. What is hybrid topology? What are the advantages of hybrid topology?
10. What are the disadvantages of hybrid topology?

[Ex. 3] Translate the following terms or phrases from English into Chinese and vice versa.

1. communication medium	1. ______
2. data link layer	2. ______
3. fibre optic cable	3. ______
4. Peer-to-Peer network	4. ______
5. radio wave	5. ______
6. *n.*地址	6. ______
7. *n.*链接	7. ______
8. *n.*互联网络	8. ______
9. *vt.*分配，散发，分发	9. ______
10. *n.*宽带	10. ______

[Ex. 4] Translate the following sentences into Chinese.

Communication Technologies Terminologies

1. Channel

Physical medium like cables over which information is exchanged is called channel. Transmission channel may be analog or digital. As the name suggests, analog channels transmit data using analog signals while digital channels transmit data using digital signals.

In popular network terminology, path over which data is sent or received is called data channel. This data channel may be a tangible medium like copper wire cables or broadcast medium like radio waves.

2. Data Transfer Rate

The speed of data transferred or received over transmission channel, measured per unit time, is called data transfer rate. The smallest unit of measurement is bits per second (bps). 1 bps means 1 bit (0 or 1) of data is transferred in 1 second.

3. Bandwidth

Data transfer rate that can be supported by a network is called its bandwidth. It is measured in bits per second (bps). Modern day networks provide bandwidth in Kbps, Mbps and Gbps. Some of the factors affecting a network's bandwidth include: network devices used, protocols used, number of users connected, network overheads like collision, errors, etc.

4. Throughput

Throughput is the actual speed with which data gets transferred over the network. Besides transmitting the actual data, network bandwidth is used for transmitting error messages, acknowledgement frames, etc.

Throughput is a better measurement of network speed, efficiency and capacity utilization rather than bandwidth.

5. Protocol

Protocol is a set of rules and regulations used by devices to communicate over the network. Just like humans, computers also need rules to ensure successful communication. Similarly, devices connected on the network need to follow some rules like when and how to transmit data, when to receive data, how to give error-free message, etc.

[Ex. 5] Fill in the blanks with the words given below.

packets	traffic	router	maximum	transmitted
port	repeater	forward	twisted	address

Router and Switch

1. Router

A router is a network layer hardware device that transmits data from one LAN to another if both networks support the same set of protocols. So a ___1___ is typically connected to at least two LANs and the internet service provider (ISP). It receives its data in the form of ___2___, which are data frames with their destination ___3___ added. Router also strengthens the signals before transmitting them. That is why it is also called ___4___.

2. Switch

Switch is a network device that connects other devices to Ethernet networks through ___5___ pair cables. It uses packet switching technique to receive, store and ___6___ data packets on the network. The switch maintains a list of network addresses of all the devices connected to it.

On receiving a packet, it checks the destination address and transmits the packet to the correct ___7___. Before forwarding, the packets are checked for collision and other network errors. The data is ___8___ in full duplex mode

Data transmission speed in switches can be double that of other network devices like hubs used for networking. This is because switch shares its ___9___ speed with all the devices connected to it. This helps in maintaining network speed even during high ___10___. In fact, higher data speeds are achieved on networks through use of multiple switches.

Online Resources

二维码	内　容
	计算机专业常用语法（8）：It 的用法
	在线阅读（1）：Transmission Media（传输介质）
	在线阅读（2）：Network Devices（网络设备）

Unit 9

Network Security

Text A

Network Security

扫码听课文

We live in an age of information. Businesses these days are more digitally advanced than ever, and as technology improves, organizations' security must be enhanced as well. Now, with many devices communicating with each other over wired, wireless, or cellular networks, network security is very important.

Network security is the process of taking physical and software preventative measures to protect the underlying networking infrastructure from unauthorized access, misuse, malfunction, modification, destruction, or improper disclosure, thereby creating a secure platform for computers, users, and programs to perform their permitted critical functions within a secure environment.

Network security is implemented by the tools you use to prevent unauthorized people or programs from accessing your networks and the devices connected to them. In essence, your computer can't be hacked if hackers can't get to it over the network.

1. Types of Network Attack

A network attack can be defined as any method, process, or means used to maliciously attempt to compromise network security. The techniques and methods used by the attacker further distinguish whether the attack is an active attack, a passive one, or some combination of the two.

1.1 Active Attacks

An active attack is a network act in which attacker attempts to make changes to data on the target or data en route to the target.

Suppose Alice wants to communicate to Bob but distance is a problem. So, Alice sends an electronic mail to Bob via a network which is not secure against attacks. There is another person, Tom, who is on the same network as Alice and Bob. Now, as the data flow is open to everyone on that network, Tom alters some portion of an authorized message to produce an unauthorized effect. For example, a message meaning "allow BOB to read confidential file X" is modified as "allow Smith to read confidential file X" (see Figure 9-1).

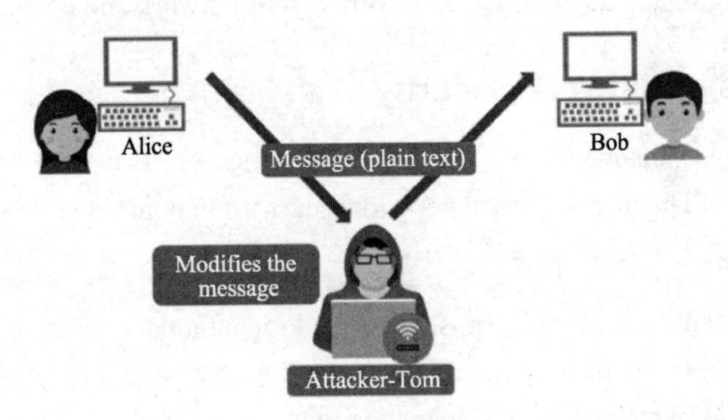

Figure 9-1 Active Attack

Active network attacks are often aggressive, blatant attacks that victims immediately become aware of when they occur. Active attacks are highly malicious in nature, often locking out users, destroying memory or files, or forcefully gaining access to a targeted system or network.

1.2 Passive Attacks

A passive attack is a network attack in which a system is monitored and sometimes scanned for open ports and vulnerabilities, but does not affect system resources.

Let's consider the example we saw earlier.

Alice sends an electronic mail to Bob via a network which is not secure against attacks. Tom, who is on the same network as Alice and Bob, monitors the data transfer that is taking place between Alice and Bob. Suppose Alice sends some sensitive information like bank account details to Bob as plain text. Tom can easily access the data and use the data for malicious purposes (see Figure 9-2).

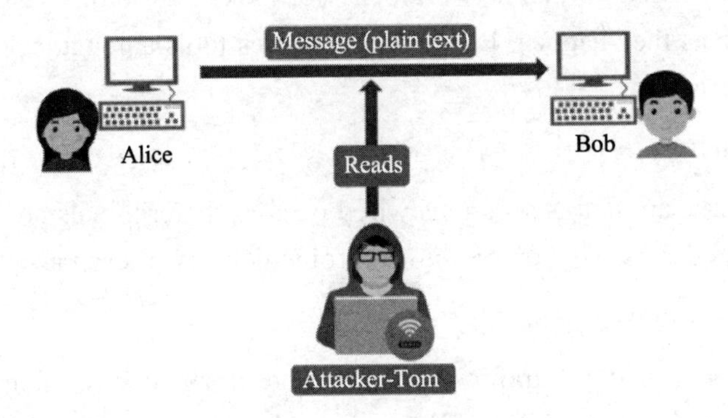

Figure 9-2 Passive Attack

The purpose of the passive attack is to gain access to the computer system or network and to collect data without detection.

So, network security includes implementing different hardware and software techniques necessary to guard underlying network architecture. With the proper network security in place, you can detect emerging threats before they infiltrate your network and compromise your data.

2. Components of Network Security

There are many components to a network security system that work together to improve your security posture. The most common network security components are discussed below.

2.1 Access Control

To keep out potential attackers, you should block unauthorized users and devices from accessing your network. Users that are permitted network access should only be able to work with the set of resources for which they've been authorized.

2.2 Application Security

Application security includes the hardware, software, and processes that can be used to track and lock down application vulnerabilities that attackers can use to infiltrate your network.

2.3 Firewalls

A firewall is a device or service that acts as a gatekeeper, deciding what enters and exits the network. They use a set of defined rules to allow or block traffic. A firewall can be hardware, software, or both.

2.4 Virtual Private Networks (VPN)

A virtual private network encrypts the connection from an endpoint to a network, often over the Internet. This way it authenticates the communication between a device and a secure network, creating a secure, encrypted "tunnel" across the open Internet.

2.5 Behavioral Analytics

You should know what normal network behavior looks like so that you can spot anomalies or network breaches as they happen. Behavioral analytics tools automatically identify activities that deviate from the norm.

2.6 Wireless Security

Wireless networks are not as secure as wired ones. Cybercriminals are increasingly targeting mobile devices and apps. So, you need to control which devices can access your network.

2.7 Intrusion Prevention System

These systems scan network traffic to identify and block attacks, often by correlating network activity signatures with databases of known attack techniques. They are some ways of implementing network security. Apart from these, you'll need a variety of software and hardware tools in your toolkit to ensure network security, those are:

- Firewalls
- Packet crafters

- Web scanners
- Packet sniffers
- Intrusion detection system
- Penetration testing software

Network security is essential for overall cyber security because network is a significant line of defense against external attack. Given that virtually all data and applications are connected to the network, robust network security protects against data breaches.

3. Security Technologies

3.1 Firewall

Firewall is a computer network security system designed to prevent unauthorized access to a private network. It can be implemented as hardware, software, or a combination of both. Firewalls are used to prevent unauthorized Internet users from accessing private networks connected to the Internet. All messages are entering or leaving the intranet pass through the firewall. The firewall examines each message and blocks those that do not meet the specified security criteria.

3.2 VPN

A VPN stands for virtual private network. It is a technology which creates a safe and an encrypted connection on the Internet from a device to a network. This type of connection helps to ensure our sensitive data is transmitted safely. It prevents our connection from eavesdropping and allows the user to access a private network securely. This technology is widely used in the corporate environments.

VPNs are also used by remote users who need to access corporate resources, consumers who want to download files and business travellers who want to access a site that is geographically restricted.

3.3 Intrusion Detection System (IDS)

An IDS is a security system which monitors the computer systems and network traffic. It analyses that traffic for possible hostile attacks originating from the outsider and also for system misuse or attacks originating from the insider. A firewall does a job of filtering the incoming traffic from the internet, the IDS in a similar way compliments the firewall security. Like the firewall protects an organization sensitive data from malicious attacks over the Internet, the intrusion detection system alerts the system administrator in the case when someone tries to break in the firewall security and tries to have access on any network in the trusted side.

3.4 Access Control

Access control is a process of selecting restrictive access to a system. It is to minimize the risk of unauthorized access to the business or organization. Users are granted access permission and certain privileges to a system and resources. Here, users must provide the credential to be granted access to a system. These credentials come in many forms such as password, keycard, the biometric reading, etc. Access control ensures security technology and access control policies to protect confidential information.

- Physical access control: This type of access control limits access to buildings, rooms, campuses, and physical IT assets.
- Logical access control: This type of access control limits connection to computer networks, system files, and data.

The more secure method for access control involves two-factor authentication. The first factor is that a user who desires access to a system must show credential and the second factor could be an access code, password, and a biometric reading.

The access control consists of two main components: authorization and authentication. Authentication is a process which verifies that someone claims to be granted access whereas an authorization provides that whether a user should be allowed to gain access to a system or denied it.

New Words

enhance	[ɪnˈhɑ:ns]	*vt.*提高，增加
cellular	[ˈseljʊlə]	*adj.*蜂窝状的
preventative	[prɪˈventətɪv]	*adj.*预防性的
protect	[prəˈt ekt]	*vt.*保护，保卫
unauthorized	[ʌnˈɔ:θəraɪzd]	*adj.*未经授权的；未经许可的
misuse	[ˌmɪsˈju:z]	*vt.*不当使用，滥用
malfunction	[ˌmælˈfʌŋkʃn]	*vi.*失灵；发生故障 *n.*故障；失灵
destruction	[dɪˈstrʌkʃn]	*n.*破坏，毁灭
improper	[ɪmˈprɒpə]	*adj.*不合适的，不正确的
disclosure	[dɪsˈkləʊʒə]	*n.*公开；泄露
platform	[ˈplætfɔ:m]	*n.*平台
critical	[ˈkrɪtɪkl]	*adj.*关键的，极重要的
hack	[hæk]	*v.*攻击，非法侵入
attack	[əˈtæk]	*n.*攻击 *vt.*侵袭；损害
maliciously	[məˈlɪʃəslɪ]	*adv.*恶意地，蓄意地
compromise	[ˈkɒmprəmaɪz]	*v.*损害
target	[ˈtɑ:gɪt]	*n.*目标，目的
alter	[ˈɔ:ltə]	*vi.*改变，修改
portion	[ˈpɔ:ʃn]	*n.*一部分

		vt.把……分成份额；分配
aggressive	[əˈgresɪv]	adj.侵略的，侵犯的；富于攻击性的
blatant	[ˈbleɪtnt]	adj.公然的；明目张胆的
victim	[ˈvɪktɪm]	n.牺牲者，受害者
forcefully	['fɔ:sfəlɪ]	adv.强有力地，强制地
affect	[əˈfekt]	vt.影响
threat	[θret]	n.威胁
infiltrate	[ˈɪnfɪltreɪt]	v.（使）渗透；（使）潜入
firewall	[ˈfaɪəwɔ:l]	n.防火墙
gatekeeper	[ˈgeɪtki:pə]	n.门卫，看门人
block	[blɒk]	vt.阻止，阻塞；限制
traffic	[ˈtræfɪk]	n.通信量，流量
		vt.用……作交换；在……通行
endpoint	['endpɔɪt]	n.端点，终点
Internet	[ˈɪntənet]	n.因特网
authenticate	[ɔ:ˈθentɪkeɪt]	vt.鉴定，使生效
anomaly	[əˈnɒməlɪ]	n.异常，反常
breach	[bri:tʃ]	vt.破坏，违反
		n.破坏；破裂；缺口
norm	[nɔ:m]	n.标准；规范；准则
cybercriminal	['saɪbəˈkrɪmɪnl]	n.计算机犯罪，网络犯罪
intrusion	[ɪnˈtru:ʒn]	n.入侵；打扰
prevention	[prɪˈvenʃn]	n.防御，预防；阻止，制止
correlate	[ˈkɒrəleɪt]	v.使互相关联；联系
packet	[ˈpækɪt]	n.信息包
		vt.包装
sniffer	['snɪfə]	n.嗅探器
penetration	[ˌpenɪˈtreɪʃn]	n.渗透，穿透
significant	[sɪgˈnɪfɪkənt]	adj.重要的，显著的
robust	[rəʊˈbʌst]	adj.强健的，结实的
examine	[ɪgˈzæmɪn]	v.检查，调查
eavesdrop	[ˈi:vzdrɒp]	v.偷听，窃听
filter	[ˈfɪltə]	vi.过滤
		n.滤波器；滤光器；滤色镜
privilege	[ˈprɪvəlɪdʒ]	n.特权

credential	[krəˈdenʃl]	*n.*凭证 *v.*提供证明书
password	[ˈpɑːswɜːd]	*n.*口令，密码
biometric	[ˌbaɪəʊˈmetrɪk]	*adj.*生物的
confidential	[ˌkɒnfɪˈdenʃl]	*adj.*秘密的，机密的
asset	[ˈæset]	*n.*资产，财产
deny	[dɪˈnaɪ]	*vt.*拒绝

Phrases

network security	网络安全
cellular network	蜂窝网络
network attack	网络攻击
attempt to	尝试，企图
active attack	主动攻击
passive attack	被动攻击
en route	在……路途中
in nature	实际上；性质上
lock out	把……关在外边，把……反锁在外
open port	开放端口
system resource	系统资源
sensitive information	敏感信息
behavioral analytic	行为分析
deviate from	背离，偏离
wireless network	无线网络
mobile device	移动设备
intrusion prevention system	入侵防御系统
penetration testing software	渗透测试软件
cyber security	网络空间安全
two-factor authentication	双因素身份验证

Abbreviations

VPN (Virtual Private Networks)	虚拟专用网
IDS (Intrusion Detection System)	入侵检测系统

Text A 参考译文

网 络 安 全

我们生活在一个信息时代。如今的业务比以往任何时候都拥有更高的数字化水平，并且随着技术的进步，组织的安全性也必须得到增强。现在，随着许多设备通过有线、无线或蜂窝网络相互通信，网络安全性非常重要。

网络安全是指采取物理和软件预防措施来保护底层网络基础结构免遭未经授权的访问、滥用、故障、修改、破坏或不当暴露的过程，从而为计算机、用户和程序创建一个安全平台，以便在安全环境中实现其允许的关键功能。

网络安全的实施就是用一些工具来防止未经授权的人员或程序访问网络及其连接的设备。从本质上讲，如果黑客无法通过网络访问计算机，则无法对计算机进行攻击。

1. 网络攻击的类型

网络攻击可以定义为用于恶意尝试危害网络安全的任何方法、过程或手段。根据攻击者使用的技术和方法，网络攻击可以进一步分为主动攻击、被动攻击或者是两者的某种组合。

1.1 主动攻击

主动攻击是一种网络行为，攻击者尝试更改目标上的数据或发往目标的数据。

假设爱丽丝想与鲍勃交流，但距离是个问题。因此，爱丽丝通过不安全的网络将电子邮件发送给鲍勃。还有另一个人——汤姆，他与爱丽丝和鲍勃在同一个网络上。现在，因为数据流向该网络上的所有人开放，汤姆更改了授权消息的某些部分以产生未授权的效果。例如，将意思为“允许鲍勃读取机密文件 X”的消息修改为“允许史密斯读取机密文件 X”。

（图略）

主动网络攻击通常是侵略性的、公然的攻击，受害者会在攻击发生时立即意识到。主动攻击本质上是非常恶意的，通常将用户锁在系统外、破坏内存或文件或强制获得对目标系统或网络的访问权限。

1.2 被动攻击

被动攻击是这样一种网络攻击：系统受到监视，有时会对其进行扫描以检查是否存在开放端口和漏洞，但不会影响系统资源。

让我们考虑一下之前看到的示例。

爱丽丝通过不安全的网络将电子邮件发送给鲍勃。与爱丽丝和鲍勃处于同一网络的汤姆监视着爱丽丝和鲍勃之间的数据传输。假设爱丽丝以纯文本形式向鲍勃发送了一些敏感信息，例如银行账户详细信息。汤姆可以轻松访问数据并将这些数据用于恶意目的。

（图略）

被动攻击的目的是获得对计算机系统或网络的访问权，并且在不被发现的情况下收集数据。

因此，网络安全包括使用必要的不同硬件和软件技术以保护底层网络体系结构。有了适当的网络安全性，就可以在新兴威胁渗透到网络并危害数据之前检测到它们。

2. 网络安全的组成部分

网络安全系统有许多组件可以协同工作以改善安全状况。下面讨论了最常见的网络安全组件。

2.1 访问控制

为了阻止潜在的攻击者，应该阻止未经授权的用户和设备访问网络。被允许访问网络的用户只能使用他们被授权使用的资源集。

2.2 应用安全

应用安全包括硬件、软件和过程，它们用来跟踪和锁定攻击者可用来渗透到网络的应用程序漏洞。

2.3 防火墙

防火墙是充当门卫的设备或服务，它决定进入和退出网络的内容。它们使用一组已定义的规则来允许或阻止流量通过。防火墙可以是硬件、软件或两者的组合。

2.4 虚拟专用网（VPN）

虚拟专用网络通常通过因特网加密从端点到网络的连接。通过这种方式，它可以验证设备和安全网络之间的通信，从而在开放的因特网上创建安全的加密“隧道”。

2.5 行为分析

你应该知道正常的网络行为是什么样的，以便可以在网络异常或发生破坏时发现它们。行为分析工具会自动识别偏离规范的活动。

2.6 无线安全

无线网络不如有线网络安全。网络犯罪分子越来越多地将目标对准移动设备和应用程序。因此，你需要控制哪些设备可以访问网络。

2.7 入侵防御系统

这些系统通常通过将网络活动特征与已知攻击技术的数据库相关联来扫描网络流入的信息，以识别和阻止攻击。它们是实现网络安全性的一些方法。除此之外，你的工具包中还需要各种软件和硬件工具来确保网络安全，这些工具包括：

- 防火墙。
- 抓包匠。
- 网络扫描仪。
- 包嗅探器。
- 入侵检测系统。
- 渗透测试软件。

网络安全对于整体网络空间安全至关重要，因为网络是抵御外部攻击的重要防线。鉴于几乎所有数据和应用程序都已连接到网络，强大的网络安全性可防止数据泄露。

3. 安全技术

3.1 防火墙

防火墙是一种计算机网络安全系统，旨在防止未经授权访问专用网络。它可以由硬件、软件或两者的组合来实现。防火墙用于防止未经授权的因特网用户访问连接到因特网的专用网络。所有消息都通过防火墙进入或离开内联网。防火墙检查每个消息，并阻止不符合指定安全标准的消息。

3.2 VPN

VPN 意为虚拟专用网。它是一种在因特网上建立从设备到网络的安全和加密连接的技术。这种类型的连接有助于确保敏感数据的安全传输。它可以防止对连接的窃听，并允许用户安全地访问专用网络。这项技术广泛用于公司环境中。

需要访问公司资源的远程用户、想要下载文件的消费者和受地理位置限制却想要访问网站的商务旅行者也都使用 VPN。

3.3 入侵检测系统（IDS）

IDS 是监视计算机系统和网络流量的安全系统。它分析流量以找到来自外部人员的可能的敌对攻击以及来自内部人员的系统滥用或攻击。防火墙负责过滤来自互联网的进入流量，IDS 以类似的方式增强了防火墙的安全性。就像防火墙可以保护组织敏感数据免受因特网上的恶意攻击一样，入侵检测系统会在有人试图破坏防火墙安全性并试图在受信任端的任何网络上访问时提醒系统管理员。

3.4 访问控制

访问控制是选择对系统的限制性访问的过程。这是为了最大程度地减少未经授权访问企业或组织的风险。用户授予被访问权限以及对系统和资源的某些特权。在此，用户必须提供凭据才能被授予访问系统的权限。这些凭据以多种形式出现，例如密码、钥匙卡、生物特征读取等。访问控制可确保安全技术和访问控制策略的实施，从而保护了机密信息。

- 物理访问控制：这种类型的访问控制限制了对建筑物、房间、校园和物理信息技术资产的访问。

- 逻辑访问控制：这种类型的访问控制限制了与计算机网络、系统文件和数据的连接。

用于访问控制的更安全的方法涉及双因素身份验证。第一个因素是希望访问系统的用户必须出示凭据，第二个因素可能是访问代码、密码和生物识别读数。

访问控制包含两个主要组件：授权和身份验证。身份验证是一个过程，该过程验证声称已被授予访问权限的用户，而授权则规定应允许用户访问系统还是拒绝用户访问。

Text B

Firewall

扫码听课文

1. What Is a Firewall

A firewall is a type of cyber security tool that is used to filter traffic on a network. Firewalls can be used to separate network nodes from external traffic sources, internal traffic sources, or even specific applications. Firewalls can be software, hardware, or cloud-based, with each type of firewall having its own unique pros and cons.

The primary goal of a firewall is to block malicious traffic requests and data packets while allowing legitimate traffic through.

2. Types of Firewalls

Firewall types can be divided into several different categories based on their general structure and method of operation. Here are eight types of firewalls.

2.1 Packet-Filtering Firewalls

As the most "basic" and oldest type of firewall architecture, packet-filtering firewalls basically create a checkpoint at a traffic router or switch. The firewall performs a simple check of the data packets coming through the router, inspecting information such as the destination and origination IP address, packet type, port number, and other surface-level information without opening up the packet to inspect its contents.

If the information packet doesn't pass the inspection, it is dropped.

The good thing about these firewalls is that they aren't very resource-intensive. This means they don't have a huge impact on system performance and are relatively simple. However, they're also relatively easy to bypass compared to firewalls with more robust inspection capabilities.

2.2 Circuit-Level Gateways

As another simplistic firewall type that is meant to quickly and easily approve or deny traffic without consuming significant computing resources, circuit-level gateways work by verifying the transmission control protocol (TCP) handshake. This TCP handshake check is designed to make sure that the session the packet is from is legitimate.

While extremely resource-efficient, these firewalls do not check the packet itself. So, if a packet holds malware, but has the right TCP handshake, it will pass right through. This is why circuit-level gateways are not enough to protect your business by themselves.

2.3 Stateful Inspection Firewalls

These firewalls combine both packet inspection technology and TCP handshake verification to create a level of protection greater than either of the previous two architectures.

However, these firewalls do put more of a strain on computing resources as well. This may slow down the transfer of legitimate packets compared to the other solutions.

2.4 Proxy Firewalls (Application-Level Gateways/Cloud Firewalls)

Proxy firewalls operate at the application layer to filter incoming traffic between your network and the traffic source — hence, the name "application-level gateway." These firewalls are delivered via a cloud-based solution or another proxy device. Rather than letting traffic connect directly, the proxy firewall first establishes a connection to the source of the traffic and inspects the incoming data packet.

This check is similar to the stateful inspection firewall in that it looks at both the packet and at the TCP handshake protocol. However, proxy firewalls may also perform deep-layer packet inspections — checking the actual contents of the information packet to verify that it contains no malware.

Once the check is complete, and the packet is approved to connect to the destination, the proxy sends it off. This creates an extra layer of separation between the "client" (the system where the packet originated) and the individual devices on your network — obscuring them to create additional anonymity and protection for your network.

If there's one drawback to proxy firewalls, it's that they can create significant slowdown because of the extra steps in the data packet transferal process.

2.5 Next-Generation Firewalls

Many of the most recently-released firewall products are being touted as "next-generation" architectures. However, there is not as much consensus on what makes a firewall truly next-gen.

Some common features of next-generation firewall architectures include deep-level packet inspection (checking the actual contents of the data packet), TCP handshake checks, and surface-level packet inspection. Next-generation firewalls may include other technologies as well, such as intrusion prevention systems (IPSs) that work to automatically stop attacks against your network.

The issue is that there is no one definition of a next-generation firewall, so it's important to verify what specific capabilities such firewalls have before investing in one.

2.6 Software Firewalls

Software firewalls include any type of firewall that is installed on a local device rather than a separate piece of hardware (or a cloud server). The big benefit of a software firewall is that it's highly useful for creating defense in depth by isolating individual network endpoints from one another.

However, maintaining individual software firewalls on different devices can be difficult and time-consuming. Furthermore, not every device on a network may be compatible with a single software firewall, which may mean having to use several different software firewalls to cover every asset.

2.7 Hardware Firewalls

Hardware firewalls use a physical appliance that acts in a manner similar to a traffic router to intercept data packets and traffic requests before they're connected to the network's servers. Physical appliance-based firewalls like this excel at perimeter security by making sure malicious traffic from outside the network is intercepted before the company's network endpoints are exposed to risk.

The major weakness of a hardware-based firewall, however, is that it is often easy for insider attacks to bypass them. Also, the actual capabilities of a hardware firewall may vary depending on the manufacturer. Some may have a more limited capacity to handle simultaneous connections than others, for example.

2.8 Cloud Firewalls

Whenever a cloud solution is used to deliver a firewall, it can be called a cloud firewall, or firewall-as-a-service (FaaS). Cloud firewalls are considered synonymous with proxy firewalls by many, since a cloud server is often used in a proxy firewall setup (though the proxy doesn't necessarily have to be on the cloud, it frequently is).

The big benefit of having cloud-based firewalls is that they are very easy to scale with your organization. As your needs grow, you can add additional capacity to the cloud server to filter larger traffic loads. Cloud firewalls, like hardware firewalls, excel at perimeter security.

3. Firewall Configuration

Firewalls are customizable. This means that you can add or remove filters based on several conditions.

3.1 IP Addresses

Each machine on the Internet is assigned a unique address called an IP address. IP addresses are 32-bit numbers, normally expressed as four "octets" in a "dotted decimal number." A typical IP address looks like this: 216.27.61.137. For example, if a certain IP address outside the company is reading too many files from a server, the firewall can block all traffic to or from that IP address.

3.2 Domain Names

Because it is hard to remember the strings of numbers that make up an IP address, and because IP addresses sometimes need to change, all servers on the Internet also have human-readable names, called domain names. A company might block all access to certain domain names, or allow access only to specific domain names.

3.3 Protocols

Some common protocols that you can set for firewall filters include:

- IP (Internet Protocol): the main delivery system for information over the Internet
- TCP (Transmission Control Protocol): used to break apart and rebuild information that travels over the Internet
- HTTP (HyperText Transfer Protocol): used for Web pages
- FTP (File Transfer Protocol): used to download and upload files
- UDP (User Datagram Protocol): used for information that requires no response, such as streaming audio and video
- ICMP (Internet Control Message Protocol): used by a router to exchange the information with other routers
- SMTP (Simple Mail Transport Protocol): used to send text-based information (e-mail)
- SNMP (Simple Network Management Protocol): used to collect system information from a remote computer
- Telnet: used to perform commands on a remote computer

A company might set up only one or two machines to handle a specific protocol and ban that protocol on all other machines.

3.4 Ports

A server machine makes its services available to the Internet using numbered ports, one for each service that is available on the server. For example, if a server machine is running a web (HTTP) server and an FTP server, the web server would typically be available on port 80, and the FTP server would be available on port 21. A company might block port 21 access on all machines but one inside the company.

3.5 Specific Words and Phrases

This can be anything. The firewall will sniff (search through) each packet of information for an exact match of the text listed in the filter. For example, you could instruct the firewall to block any packet with the word "X-rated" in it. The key here is that it has to be an exact match. The "X-rated" filter would not catch "X rated" (no hyphen). But you can include as many words, phrases and variations of them as you need.

Some operating systems come with a firewall built in. Otherwise, a software firewall can be installed on the computer in your home that has an Internet connection. This computer is considered a gateway because it provides the only point of access between your home network and the Internet.

With a hardware firewall, the firewall unit itself is normally the gateway. Computers in your home network connect to the router, which in turn is connected to either a cable or DSL modem. You configure the router via a web-based interface that you reach through the browser on your computer. You can then set any filters or additional information.

New Words

malicious	[məˈlɪʃəs]	*adj*.恶意的，有敌意的
legitimate	[lɪˈdʒɪtɪmət]	*adj*.合法的，合理的
checkpoint	[ˈtʃekpɔɪnt]	*n*.检查点，检查站
destination	[ˌdestɪˈneɪʃn]	*n*.目的，目标；目的地
origination	[əˌrɪdʒɪˈneɪʃən]	*n*.起源，来源，起始
port	[pɔ:t]	*n*.端口
bypass	[ˈbaɪpɑ:s]	*v*.绕开，避开 *n*.旁道，支路
gateway	[ˈgeɪtweɪ]	*n*.网关
handshake	[ˈhændʃeɪk]	*n*.握手信号
session	[ˈseʃn]	*n*.会话，会话期
malware	[ˈmælweə]	*n*.恶意软件，流氓软件
proxy	[ˈprɒksɪ]	*n*.代理；代理服务器
layer	[ˈleɪə]	*n*.层，层次 *vi*.分成层次
cloud-based	[klaʊd-beɪst]	*adj*.基于云的
actual	[ˈæktʃuəl]	*adj*.真实的，实际的
obscure	[əbˈskjʊə]	*vt*.使……模糊不清，掩盖；隐藏
slowdown	[ˈsləʊdaʊn]	*n*.减速，减缓
generation	[ˌdʒenəˈreɪʃn]	*n*.代，时代；产生
consensus	[kənˈsensəs]	*n*.一致，一致同意
isolate	[ˈaɪsəleɪt]	*v*.隔离，分离；孤立
cover	[ˈkʌvə]	*v*.覆盖，遮盖
risk	[rɪsk]	*n*.风险，危险，冒险 *vt*.冒……的危险；使……冒风险

manufacturer	[ˌmænjʊˈfæktʃərə]	*n.*制造商，制造厂
handle	[ˈhændl]	*v.*处理，操作，操纵 *n.*句柄；手柄
synonymous	[sɪˈnɒnɪməs]	*adj.*同义的，类义的
load	[ləʊd]	*n.*负载，负荷，装载；工作量 *v.*装入，装上
octet	[ɒkˈtet]	*n.*八位字节
string	[strɪŋ]	*n.*串，字符串
response	[rɪˈspɒns]	*n.*反应，响应；回答
text-based	[tekst-beɪst]	*adj.*基于文本的
sniff	[snɪf]	*v.*嗅探
match	[mætʃ]	*v.*匹配 *n.*比赛；相配的人（或物）
hyphen	[ˈhaɪfn]	*n.*连字符，连字号
modem	[ˈməʊdem]	*n.*调制解调器

Phrases

data packet	数据包
packet-filtering firewall	包过滤防火墙
port number	端口号
circuit-level gateway	电路级网关
proxy firewall	代理防火墙
application layer	应用层
stateful inspection firewall	状态检测防火墙
being touted as	被吹捧为
local device	本地设备
cloud server	云服务器
be compatible with	与……兼容，与……一致
be exposed to	暴露于，暴露在；面临
domain name	域名

Abbreviations

TCP (Transmission Control Protocol)	传输控制协议
IPSs (Intrusion Prevention Systems)	入侵防御系统
FaaS (Firewall-as-a-Service)	防火墙即服务

FTP (File Transfer Protocol)	文件传输协议
UDP (User Datagram Protocol)	用户数据报协议
ICMP (Internet Control Message Protocol)	因特网控制消息协议
SMTP (Simple Mail Transport Protocol)	简单邮件传输协议
SNMP (Simple Network Management Protocol)	简单网络管理协议
DSL (Digital Subscriber Line)	数字用户线路

Text B 参考译文

防　火　墙

1. 什么是防火墙

防火墙是一种网络空间安全工具，用于过滤网络流量。防火墙可用于将网络节点与外部流量源、内部流量源甚至特定应用程序分开。防火墙可以是基于软件的、基于硬件的或基于云的，每类防火墙都有其各自的优缺点。

防火墙的主要目标是阻止恶意数据流请求和数据包，同时允许合法数据流通过。

2. 防火墙的类型

根据防火墙的一般结构和操作方法，可以将其分为几种不同的类别。下面介绍 8 种类型的防火墙。

2.1　包过滤防火墙

作为最“基础”和最古老的防火墙体系结构，包过滤防火墙基本上在流路由器或交换机上创建检查点。防火墙对通过路由器的数据包进行简单检查，检查诸如目的地和始发 IP 地址、包类型、端口号以及其他表层信息，而无须打开包来检查其内容。

如果信息包未通过检查，则会被丢弃。

这些防火墙的好处是它们并不占用大量资源。这意味着它们不会对系统性能产生重大影响，而且相对简单。但是，与具有更强大的检查功能的防火墙相比，它们也相对容易被绕开。

2.2　电路级网关

作为另一种简单的防火墙类型，电路级网关旨在不消耗大量计算资源而快速、轻松地批准或拒绝数据流，它通过验证传输控制协议（TCP）握手信号来工作。该 TCP 握手检查旨在确保会话的数据包是合法的。

尽管这些防火墙非常节省资源，但它们不会检查数据包本身。因此，如果数据包中包含恶意软件，但具有正确的 TCP 握手信号，则它将直接通过。这就是电路级网关不足以单独保护业务的原因。

2.3 状态检查防火墙

这些防火墙将数据包检查技术和 TCP 握手验证相结合，以创建比前两种体系结构都更高的保护级别。

但是，这些防火墙也确实给计算资源带来了更大的压力。与其他解决方案相比，这可能会减慢合法数据包的传输。

2.4 代理防火墙（应用级网关/云防火墙）

代理防火墙在应用层运行，以过滤网络和流源之间的进入流——因此叫做“应用级网关”。这些防火墙是通过基于云的解决方案或其他代理设备交付的。代理防火墙不让数据流直接连接，而是先建立与流源的连接并检查传入的数据包。

该检查与状态检查防火墙相似，因为它同时检查数据包和 TCP 握手协议。但是，代理防火墙也可以执行深层数据包检测——检查信息数据包的实际内容，以确认它不包含恶意软件。

一旦检查完成，并且数据包被批准连接到目标，则代理将其发送出去。这在“客户端”（数据包的始发系统）和网络上的各个设备之间形成了额外的隔离层——将其隐藏从而为网络增加了匿名性并提供了保护。

如果说代理防火墙有一个缺点的话，那就是数据包传输过程中的额外步骤会大大降低传输的速度。

2.5 下一代防火墙

许多最新发布的防火墙产品被吹捧为“下一代”体系结构。但是，关于什么是真正的下一代防火墙，尚未达成共识。

下一代防火墙体系结构的一些常见功能包括深层包检测（检查数据包的实际内容）、TCP 握手检查和表层包检查。下一代防火墙也可能包括其他技术，例如入侵防御系统（IPS），该技术可自动阻止针对网络的攻击。

问题在于，对下一代防火墙没有一个定义，因此在投资这种防火墙之前，必须先验证此类防火墙具有哪些特定功能，这一点很重要。

2.6 软件防火墙

软件防火墙包括安装在本地设备而不是单独的硬件（或云服务器）上的任何类型的防火墙。软件防火墙的最大好处是将各个网络端点彼此隔离，它对于深度防御非常有用。

但是，在不同的设备上维护单独的软件防火墙可能是困难且耗时的。此外，并非网络上的每个设备都可能与单个软件防火墙兼容，这可能意味着必须使用多个不同的软件防火墙来覆盖每个设备。

2.7 硬件防火墙

硬件防火墙使用一种物理设备，该物理设备的行为类似于流路由器，以在将数据包和流量请求连接到网络服务器之前对其进行拦截。这样的基于物理设备的防火墙通过确保在公司的网络终节点遭受风险之前拦截来自网络外部的恶意流，从而在外围安全方面表现出色。

但是，基于硬件的防火墙的主要缺点是内部攻击者通常很容易绕过它们。另外，硬件防火墙的实际功能可能会因制造商而异。例如，某些服务器可能处理同时连接的能力有限。

2.8 云防火墙

使用云解决方案交付的防火墙都可以称为云防火墙或防火墙即服务（FaaS）。许多人认为云防火墙与代理防火墙是同义词，因为云服务器通常用于代理防火墙设置中（尽管代理防火墙并非一定要在云上，但它经常在云上）。

拥有基于云的防火墙的最大好处是，它们很容易跟随组织来扩展。随着需求的增长，可以向云服务器添加更多容量以过滤更大的流负载。像硬件防火墙一样，云防火墙在外围安全方面也很出色。

3. 防火墙配置

防火墙是可定制的。这意味着可以根据多种条件添加或删除过滤器。

3.1 IP 地址

因特网上的每台计算机都分配有唯一的地址，称为 IP 地址。IP 地址是 32 位数字，通常用“点分十进制数字”表示为 4 个“八位字节”。典型的 IP 地址如下所示：216.27.61.137。例如，如果公司外部的某个 IP 地址正在从服务器读取太多文件，则防火墙可能会阻止与该 IP 地址之间的所有通信。

3.2 域名

由于很难记住组成 IP 地址的数字串，并且由于 IP 地址有时需要更改，所以因特网上的所有服务器也都具有人类可读的名称，称为域名。公司可能会阻止对某些域名的所有访问，或者仅允许对特定域名的访问。

3.3 协议

可以设置防火墙过滤器的一些常见协议包括：

- IP（网际协议）：因特网上主要的信息传递系统。
- TCP（传输控制协议）：用于分解和重建通过因特网传输的信息。

- HTTP（超文本传输协议）：用于网页。
- FTP（文件传输协议）：用于下载和上传文件。
- UDP（用户数据报协议）：用于不需要响应的信息，例如流音频和视频。
- ICMP（因特网控制消息协议）：路由器用来与其他路由器交换信息。
- SMTP（简单邮件传输协议）：用于发送基于文本的信息（电子邮件）。
- SNMP（简单网络管理协议）：用于从远程计算机收集系统信息。
- Telnet：用于在远程计算机上执行命令。

公司可能只设置一台或两台计算机来处理特定协议，并在其他所有计算机上禁止该协议。

3.4 端口

服务器计算机使用编号的端口将其服务提供给因特网，而服务器上的每种服务均使用一个端口。例如，如果服务器计算机正在运行 Web（HTTP）服务器和 FTP 服务器，则该 Web 服务器通常在端口 80 上可用，而 FTP 服务器在端口 21 上可用。公司可能会阻止全部机器对端口 21 的访问，但内部有一台机器可以使用。

3.5 特定词和短语

这可以是任何东西。防火墙将嗅探（搜索）每个信息包，以查找与过滤器中列出的文本完全匹配的信息。例如，你可以指示防火墙阻止其中带有单词“X-rated”的任何数据包。关键是必须完全匹配。“X-rated”过滤器不会捕获“X rated”（无连字符）。但是可以根据需要包含尽可能多的单词、短语和它们的变体形式。

某些操作系统内置了防火墙。当然也可以在与因特网连接的家庭计算机上安装软件防火墙。该计算机被认为是网关，因为它提供了家庭网络和因特网之间的唯一访问点。

对于硬件防火墙，防火墙单元本身通常就是网关。家庭网络中的计算机连接到路由器，该路由器又连接到电缆或 DSL 调制解调器。你可以用计算机上的浏览器通过基于 Web 的界面来配置路由器。然后，你可以设置任何过滤器或附加信息。

Exercises

[Ex. 1] Answer the following questions according to Text A.

1. What is network security?
2. What can a network attack defined as?
3. What is an active attack?
4. What is a passive attack?
5. What are the most common network security components?

6. What does application security include?
7. What does VPN stand for? What is it?
8. What is an IDS? What does it do?
9. What is access control?
10. What does the access control consist of? What are they respectively?

[Ex.2] Fill in the following blanks according to Text B.

1. A firewall is a type of cyber security tool that is used to ____________________. Firewalls can be used to ____________________ from external traffic sources, ____________________, or even ____________________.
2. The firewall performs a simple check of the data packets coming through ______________, inspecting information such as the destination and origination ___________, ______________, ______________ and other surface-level information without ______________to inspect its contents.
3. These firewalls combine both ______________ and ______________to create a level of protection greater than either of the previous two architectures.
4. Proxy firewalls operate at ______________ to filter incoming traffic between ______________ and ______________ — hence, the name "application-level gateway". These firewalls are delivered via ______________or another proxy device.
5. Software firewalls include any type of firewall that is installed ______________ rather than ______________ (or a cloud server). The big benefit of a software firewall is that it's highly useful for ______________ by isolating ______________ from one another.
6. The major weakness of a hardware-based firewall, however, is that it is often easy for ______________ to bypass them. Also, the actual capabilities of a hardware firewall may vary depending on ______________. Some may have ______________ to handle simultaneous
7. The big benefit of having cloud-based firewalls is that they are very easy ________________. As your needs grow, you can ________________ to the cloud server to ________________.
8. Each machine on the Internet is assigned ________________ called an IP address. IP addresses are ________________, normally expressed as ________________ in a "dotted decimal number".
9. Some common protocols that you can set for firewall filters include: __________, __________, __________, FTP, UDP, __________, SMTP, __________, Telnet.
10. Computers in your home network connect to the router, which in turn is connected to either __________ or DSL modem. You configure the router via ____________ that

you reach through ____________ on your computer. You can then set ____________ or additional information.

[Ex. 3] Translate the following terms or phrases from English into Chinese and vice versa.

1. active attack	1. ____________
2. cyber security	2. ____________
3. intrusion prevention system	3. ____________
4. passive attack	4. ____________
5. sensitive information	5. ____________
6. *vi.*改变，修改	6. ____________
7. *adj.*秘密的.机密的	7. ____________
8. *v.*使互相关联；联系	8. ____________
9. *n.*入侵；打扰	9. ____________
10. *n.*渗透，穿透	10. ____________

[Ex. 4] Translate the following sentences into Chinese.

The Hacker Toolbox

The main resource hackers rely upon, apart from their own ingenuity, is computer code. While there is a large community of hackers on the Internet, only a relatively small number of hackers actually program code. Many hackers seek out and download code written by other people. There are thousands of different programs hackers use to explore computers and networks. These programs give hackers a lot of power over innocent users and organizations—once a skilled hacker knows how a system works, he can design programs that exploit it.

Malicious hackers use programs to:

- Log keystrokes: Some programs allow hackers to review every keystroke a computer user makes. Once installed on a victim's computer, the programs record each keystroke, giving the hacker everything he needs to infiltrate a system or even steal someone's identity.
- Hack passwords: There are many ways to hack someone's password, from educated guesses to simple algorithms that generate combinations of letters, numbers and symbols. The trial and error method of hacking passwords is called a brute force attack, meaning the hacker tries to generate every possible combination to gain access. Another way to hack passwords is to use a dictionary attack, a program that inserts common words into password fields.
- Infect a computer or system with a virus: Computer viruses are programs designed to

duplicate themselves and cause problems ranging from crashing a computer to wiping out everything on a system's hard drive. A hacker might install a virus by infiltrating a system, but it's much more common for hackers to create simple viruses and send them out to potential victims via email, instant messages, web sites with downloadable content or peer-to-peer networks.

- Gain backdoor access: Similar to hacking passwords, some hackers create programs that search for unprotected pathways into network systems and computers. In the early days of the Internet, many computer systems had limited security, making it possible for a hacker to find a pathway into the system without a username or password. Another way a hacker might gain backdoor access is to infect a computer or system with a Trojan horse.
- Create zombie computers: A zombie computer is a computer that a hacker can use to send spam or commit distributed denial of service (DDoS) attacks. After a victim executes seemingly innocent code, a connection opens between his computer and the hacker's system. The hacker can secretly control the victim's computer, using it to commit crimes or spread spam.
- Spy on email: hackers have created code that lets them intercept and read email messages . Today, most email programs useencryption formulas so complex that even if a hacker intercepts the message, he won't be able to read it.

[Ex. 5] Fill in the blanks with the words given below.

threats	signatures	suspicious	involves	wireless
malicious	methods	intrusion	monitor	vulnerabilities

Intrusion Prevention Systems

Intrusion prevention systems have various ways of detecting malicious activity, however the two predominant __1__ are signature-based detection and statistical anomaly-based detection. The signature-based detection method used by intrusion prevention systems __2__ a dictionary of uniquely identifiable __3__ located in the code of each exploit. There are two types of signature-based detection methods for intrusion prevention systems as well: exploit-facing and vulnerability-facing. Exploit-facing methods detect __4__ activity based on common attack patterns, whereas vulnerability-facing methods attempt to detect malicious activity by identifying specific __5__. On the other hand, intrusion prevention systems that rely on statistical anomaly-based detection randomly sample network traffic and then compare the samples to a predetermined baseline performance level.

There are four common types of intrusion prevention systems. The first type of __6__ prevention system is called a network-based intrusion prevention system (NIPS). This type of intrusion prevention system has the ability to __7__ the whole network and look for suspicious traffic by reviewing protocol activity. In contrast, wireless intrusion prevention systems (WIPS) only monitor wireless networks for suspicious activity by reviewing ___8___ networking protocols. A third type of intrusion prevention system is called network behavior analysis (NBA). Network behavior analysis looks at network traffic in an effort to locate ___9___ that cause unusual traffic flows, including distributed denial of service (DDoS) attacks and policy violations. The last common type of intrusion prevention system is host-based intrusion prevention systems (HIPS). A host-based intrusion prevention systems is an installed software package that looks into ___10___ activity that occurs within a single host.

Online Resources

二维码	内　容
	计算机专业常用语法（9）：被动语态
	在线阅读（1）：8 Different Types of Malware （8 种不同类型的恶意软件）
	在线阅读（2）：12 Ways to Secure Your Computer from Hackers （保护计算机免受黑客攻击的 12 种方法）

Unit 10

Cloud Computing and Cloud Storage

Text A

Cloud Computing

扫码听课文

1. What Is Cloud Computing

The internet is changing the way we conduct business and interact. Traditionally, hardware and software are fully contained on a user's computer. This means that you access your data and programs exclusively within your own computer.

Cloud computing allows you to access your data and programs outside of your own computing environment. Rather than storing your data and software on your personal computer or server, it is stored in "the cloud". This could include applications, databases, email and file services.

A common factor to consider when using cloud computing is renting versus buying. Essentially, you rent capacity (server space or access to software) from a cloud service provider, and connect over the internet. Instead of buying your own IT requirements, you are renting from a service provider, paying for only the resources you use.

Cloud computing has four models in terms of different access and security options. Before you move your data into the cloud, you will need to consider which model works best for your business and data needs.

1.1 Private Cloud

A private cloud is where the services and infrastructures are maintained and managed by you or a third party. This option reduces the potential security and control risks, and will suit you if your data and applications are a core part of your business and you need a higher degree of security or have sensitive data requirements.

1.2 Community Cloud

A community cloud exists where several organizations share access to a private cloud, with similar security considerations. For example, a series of franchises have their own public clouds, but they are hosted remotely in a private environment.

1.3 Public Cloud

A public cloud is where the services are stored off-site and accessed over the internet. The storage is managed by an external organization such as Google or Microsoft. This service offers the greatest level of flexibility and cost saving, however, it is more vulnerable than private clouds.

1.4 Hybrid Cloud

A hybrid cloud model takes advantages of both public and private cloud services. By spreading your options across different cloud models, you gain the benefits of each model.

For example, you could use a public cloud for your emails to save on large storage costs, while keeping your highly sensitive data safe and secure behind your firewall in a private cloud.

2. How Cloud Computing Works

There are three main types of cloud computing service models available, commonly known as software as a service (SaaS), infrastructure as a service (IaaS) and platform as a service (PaaS).

Depending on your needs, your business could use one of these service models, or a mixture of the three.

2.1 Software as a Service (SaaS)

SaaS is the most common form of cloud computing for small businesses. You can access internet-hosted software applications using a browser, rather than traditional applications stored on your own PC or server. The software application host is responsible for controlling and maintaining the application, including software updates and settings. You, as a user, have limited control over the application and configuration settings. A typical example of a SaaS is a web-based mail service or customer relationship management system.

2.2 Infrastructure as a Service (IaaS)

IaaS typically means buying or renting your computer power and disk space from an external service provider. This option allows you to access through a private network or over the internet. The service provider maintains the computer hardware including CPU processing, memory, data storage and network connectivity.

2.3 Platform as a Service (PaaS)

PaaS can be described as a crossover of both SaaS and IaaS. Essentially you rent the hardware, operating systems, storage and network capacity that IaaS provides, as well as the software servers and application environments. PaaS offers you more control over the technical aspects of your computing setup and the ability to customize to suit your needs.

3. Benefits of Cloud Computing

Cloud computing offers your business many benefits. It allows you the flexibility of connecting to your business anywhere any time. With the growing number of web-enabled devices used in today's business environment (e.g. smartphones, tablets), it is even easier access to your data. There are many benefits to moving your business to the cloud.

3.1 Reduced IT Costs

Moving to cloud computing may reduce the cost of managing and maintaining your IT systems. Rather than purchasing expensive systems and equipment for your business, you can reduce your costs by using the resources of your cloud computing service provider. You may be able to reduce your operating costs because:

- The cost of system upgrades, new hardware and software may be included in your contract.
- You no longer need to pay wages for expert staff.
- Your energy consumption costs may be reduced.
- There are fewer time delays.

3.2 Scalability

Your business can scale up or scale down your operation and storage needs quickly to suit your situation, allowing flexibility as your needs change. Rather than purchasing and installing expensive upgrades yourself, your cloud computer service provider can handle this for you. Using the cloud frees up your time so you can get on with running your business.

3.3 Business Continuity

Protecting your data and systems is an important part of business continuity planning. Whether you experience a natural disaster, power failure or other crisis, having your data stored in the cloud ensures it is backed up and protected in a secure and safe location. Being able to access your data again quickly allows you to conduct business as usual, minimize any downtime and loss of productivity.

3.4 Collaboration Efficiency

Collaboration in a cloud environment gives your business the ability to communicate and share more easily outside of the traditional methods. If you are working on a project across different locations, you could use cloud computing to give employees, contractors and third parties access to the same files. You could also choose a cloud computing model that makes it easy for you to share your records with your advisers. For example, you could share accounting records with your accountant or financial adviser in a quick and secure way.

3.5 Flexibility of Work Practices

Cloud computing allows employees to be more flexible in their work practices. For example, you have the ability to access data from home, on holiday, or via the commute to and from work. If you need access to your data while you are off-site, you can connect to your virtual office, quickly and easily.

3.6 Access to Automatic Updates

Access to automatic updates for your IT requirements may be included in your service fee. Your system will regularly be updated with the latest technology by your cloud computing service provider. This could include up-to-date versions of software, as well as upgrades to servers and computer processing power.

4 Risks of Cloud Computing

Before considering cloud computing technology, it is important to understand the risks involved when moving your business into the cloud. You should carry out a risk assessment before any control is handed over to a service provider.

4.1 Privacy Agreement and Service Level Agreement

You will need to have suitable agreements in place with your service providers before services commence. This will safeguard you against certain risks and also outline the responsibilities of each party in the form of a service level agreement (SLA). You should read the SLA and ensure that you understand what you are agreeing to before you sign. Make sure that you understand the responsibilities of the service provider, as well as your own obligations.

4.2 Security and Data Protection

You must consider how your data will be stored and secured when outsourcing to a third party. This should be outlined in the agreement with your service provider, and must control security risks. It must cover who has access to the data and the security measures must be in place to protect your data.

4.3 Location of Data

Cloud computing service providers can be located at home or abroad. Before committing, you should investigate where your data is being stored and which privacy and security laws will apply to the data.

New Words

interact	[ˌɪntər'ækt]	*v.*交互，交流，沟通；相互作用；互相影响
analogy	[ə'nælədʒɪ]	*n.*类似，相似；类推
option	['ɒpʃn]	*n.*选项；选择权
potential	[pə'tenʃl]	*adj.*潜在的，有可能的
control	[kən'trəʊl]	*vt.*控制，管理 *n.*（键盘上的）控制键

suit	[su:t]	*vt.*适合于 *vi.*合适，相称
degree	[dɪˈgri:]	*n.*程度；度数
off-site	['ɔ:f'sɑɪt]	*adv.*站外；装置外
flexibility	[ˌfleksə'bɪlɪtɪ]	*n.*柔韧性，机动性，灵活性
spread	[spred]	*v.*伸开，展开；（使）传播，（使）散布
set	[set]	*vt.*设置，放置，安置
connectivity	[ˌkɒnekˈtɪvɪtɪ]	*n.*连通性
setup	['setʌp]	*n.*设置，安装
continuity	[ˌkɒntɪˈnju:ətɪ]	*n.*连续性，继续
disaster	[dɪˈzɑ:stə]	*n.*灾难
crisis	[ˈkraɪsɪs]	*n.*危机；决定性时刻，紧要关头
productivity	[ˌprɒdʌkˈtɪvɪtɪ]	*n.*生产率，生产力
collaboration	[kəˌlæbəˈreɪʃn]	*n.*合作，协作
project	[ˈprɒdʒekt]	*vt.*项目，工程，计划
record	[ˈrekɔ:d]	*n.*记录，记载；档案
accountant	[əˈkaʊntənt]	*n.*会计人员，会计师
suitable	[ˈsu:təbl]	*adj.*合适的，适当的
responsibility	[rɪˌspɒnsəˈbɪlɪtɪ]	*n.*责任，职责
obligation	[ˌɒblɪˈgeɪʃn]	*n.*义务，责任
outsource	[ˈaʊtsɔ:s]	*vt.*外购，外包
location	[ləʊˈkeɪʃn]	*n.*位置，场所

Phrases

cloud computing	云计算
personal computer	个人计算机
server space	服务器空间
cloud service provider	云服务提供商，云服务提供者
in terms of	根据；就……而言
private cloud	私有云，专用云
third party	第三方
community cloud	社区云
public cloud	公共云
cost saving	节省成本
hybrid cloud	混合云

customer relationship management	客户关系管理
disk space	磁盘空间
be described as	被描述为
operating cost	运营成本
energy consumption cost	电力消耗成本，能源消耗成本
natural disaster	自然灾难
financial adviser	财务顾问
virtual office	虚拟办公室

Abbreviations

SaaS (Software as a Service)	软件即服务
IaaS (Infrastructure as a Service)	基础设施即服务
PaaS (Platform as a Service)	平台即服务
SLA (Service-Level Agreement)	服务水平协议，服务级别协议

Text A 参考译文

云 计 算

1. 什么是云计算

互联网正在改变我们开展业务和交互的方式。传统上，硬件和软件完全包含在用户的计算机中。这意味着你只能访问自己计算机上的数据和程序。

云计算使你可以访问自己计算环境之外的数据和程序。它不是将数据和软件存储在个人计算机或服务器上，而是存储在"云"中。这可能包括应用程序、数据库、电子邮件和文件服务。

使用云计算常需要考虑租赁还是购买。本质上，你是从云服务提供商租用容量（服务器空间或对软件的访问权），然后通过互联网连接。你无须购买自己的 IT 设备，而是从服务提供商处租用，仅须为所使用的资源付费。

就不同的访问和安全选项而言，云计算有 4 种模式。将数据移到云中之前，你需要考虑哪种模式最适合你的业务和数据需求。

1.1 私有云

私有云是你或第三方维护和管理服务和基础设施的地方。如果你的数据和应用程序是业务的核心部分，并且你需要更高级别的安全性或对数据要求很在意，该选项可降低潜在的安全和控制风险，它将非常适合你。

1.2 社区云

社区云就是多个组织出于相似的安全考虑，共享对私有云的访问。例如，某些专营商有自己的公共云，但是它们是在私有环境中远程托管的。

1.3 公共云

公共云是将服务存储在异地并通过互联网访问的地方。存储由外部组织（例如谷歌或微软）管理。该服务提供了最大程度的灵活性，也节省成本，但是，它比私有云更容易受到攻击。

1.4 混合云

混合云模式利用了公共云服务和私有云服务的优势。通过将选择分散在不同的云模式中，可以获得每种模式的好处。

例如，可以将公共云用于电子邮件以节省大量存储成本，同时把高度敏感的数据保存在防火墙后面的私有云中，使其安全。

2. 云计算如何工作

共有三种主要类型的云计算服务模式，通常称为软件即服务（SaaS），基础架构即服务（IaaS）和平台即服务（PaaS）。

根据需求，企业可以使用其中一种服务模式，也可以混合使用这三种服务模式。

2.1 软件即服务（SaaS）

SaaS 是小型企业最常见的云计算形式。你可以使用浏览器访问互联网托管的软件应用程序，而不是使用存储在自己的 PC 或服务器上的传统的应用程序。软件应用程序主机负责控制和维护应用程序，包括软件更新和设置。作为用户，对应用程序和配置设置的控制有限。SaaS 的典型示例是基于 Web 的邮件服务或客户关系管理系统。

2.2 基础设施即服务（IaaS）

IaaS 通常意味着从外部服务提供商那里为计算机购买或租用性能和磁盘空间。该选项使你可以通过专用网络或互联网访问资源。服务提供商维护计算机硬件（包括 CPU 处理、内存、数据存储）和网络连接。

2.3 平台即服务（PaaS）

PaaS 可以描述为 SaaS 和 IaaS 的交叉。本质上，你租用了 IaaS 提供的硬件、操作系统、存储和网络容量以及软件服务器和应用程序环境。 PaaS 使你可以更好地控制计算设置的技术方面，并可以进行定制以适应需求。

3. 云计算的好处

云计算为业务带来许多好处。它使你可以随时随地灵活地连接业务。随着当今商业环境中使用的支持网络的设备（例如智能手机、平板电脑）的数量不断增加，访问数据变得更加容易。将业务迁移到云有很多好处。

3.1 降低 IT 成本

迁移到云计算可以减少管理和维护 IT 系统的成本。可以通过使用云计算服务提供商的资源来降低成本，而不必为企业购买昂贵的系统和设备。运营成本可能能够降低，因为：

- 合同中可能包含系统升级、新的硬件和软件的成本。
- 不再需要为专业人员支付工资。
- 能源消耗成本可能会减少。
- 时间延迟更少。

3.2 可扩展性

企业可以根据情况快速扩展或缩减运行和存储需求，并根据需求的变化提供灵活性。云计算机服务提供商可以解决此问题，而升级时不必自己购买和安装昂贵的设备。使用云可以节省时间，因此可以继续经营自己的业务。

3.3 业务连续性

保护数据和系统是业务连续性计划的重要组成部分。无论遇到自然灾害、电源故障还是其他危机，将数据存储在云中都可以确保在安全的位置备份和保护数据。你能够再次快速访问数据使业务能够照常开展，从而最大程度地缩短停机时间和减少生产力损失。

3.4 协作效率

在云环境中的协作使企业能够在传统方法之外更轻松地进行通信和共享。如果你正从事的项目跨越不同位置，则使用云计算能让员工、承包商和第三方访问相同的文件。还可以选择一种云计算模式，使你可以轻松地与顾问共享记录。例如，可以快速、安全地与会计或财务顾问共享会计记录。

3.5 工作方式的灵活性

云计算使员工可以更加灵活地开展工作。例如，可以在家中、度假时或上下班途中访问数据。如果需要不在现场时访问数据，则可以快速、轻松地连接到虚拟办公室。

3.6 访问自动更新

服务费中可能包含对 IT 要求的自动更新的访问权限。云服务提供商将定期使用最新技术对系统进行更新。这可能包括软件版本更新以及服务器和计算机处理能力的升级。

4. 云计算的风险

在考虑云计算技术之前，重要的是要了解将业务迁移到云中时所涉及的风险。在将任何控制权移交给服务提供商之前，都应该进行风险评估。

4.1 隐私协议和服务水平协议

服务开始之前，你需要与服务提供商达成适当的协议。这将保护你免受某些风险的侵害，并以服务水平协议（SLA）的形式概述各方的责任。你应该阅读SLA，并确保在签署前了解所同意的内容。确保了解服务提供商的职责以及你自己的义务。

4.2 安全性和数据保护

当外包给第三方时，必须考虑如何存储和保护数据。在与服务提供商的协议中应该对此进行概述，并且必须控制安全风险。它必须涵盖有权访问数据的人员，并且必须采取安全措施来保护数据。

4.3 数据位置

云计算服务提供商可以位于国内或国外。在提交之前，应该调查数据的存储位置以及哪些隐私和安全法将适用于该数据。

Text B

Cloud Storage

扫码听课文

1. What Is Cloud Storage

Cloud storage is file storage in the cloud (online). Instead of keeping your files on your local hard drive, external hard drive, or flash drive, you can save them online.

There are multiple reasons to use cloud storage services. Maybe your local hard drives are running low on disk space, in which case you can use the cloud as extra storage. If you want to stream your music collection from anywhere, access your work files at home, easily share vacation videos, etc., you can upload your files online to a cloud storage service. Another reason to use cloud storage is if you want to keep important files secure behind a password and encryption.

In short, cloud storage is helpful not only when it comes to backup but also for security and the ability to easily share files with others or access them yourself from anywhere: your phone, tablet, or another computer.

2. How Cloud Storage Works

When you upload a file to the internet and that file is there for an extended period of time, it's considered cloud storage. The simplest type of cloud storage is uploading something to a server and having the ability to retrieve it again should you want to.

A reputable cloud storage service protects the files behind encryption and requires you to enter a password in order to be able to access the files. Most of the time, the cloud storage account can be protected behind two-factor authentication, too, so that anyone wanting access to your files has to know not only the password but another code sent to your phone upon the login request.

Most cloud storage services let you upload all types of files, videos, pictures, documents, music, or anything else. However, some are limited to accept only certain kinds of files, such as only images or music. Cloud storage services are usually fairly clear about what's allowed and what isn't.

Different cloud storage services let you upload files to your online account through different methods. Some support in-browser uploads only, meaning that you have to log in to the cloud storage service's website to upload your data, but most have desktop applications that make uploading files easier by a simple drag-and-drop into the service's dedicated folder. Most also support uploading images and videos from your phone.

Less common are torrent cloud storage services that are online torrent clients that not only let you download torrents from your browser but also store your files in your online account to stream or download later.

Once your files are stored online, depending on how the service works, the features you get might include the ability to stream videos and music, access the files from your mobile device, easily share the files with others through a special shared link, download the files back to your computer, delete them to free up space in your account, encrypt them so that even the service can not see them, and more.

3. Cloud Storage vs. Cloud Backup

Cloud storage and cloud backup are easily confused. They both work similarly and have a similar end result, that is the files are stored online. But there are two completely different reasons to use these services, and knowing how they differ is important so that you know which one to choose for your own situation.

Cloud storage is a selective backup procedure where you choose which files to store online, and then you send them to your online account. When you delete a file that you backed up online on your computer, the file is still in your cloud storage account because it isn't actually tied to your computer anymore, it's just a single file that you uploaded online.

Cloud backup is that you install a program on your computer and tell it to keep specific files backed up online. Going a step further than cloud storage, a backup service will also upload any changes you make to the file so that the current version is always stored online. In other words, if you delete a file from your computer, it might also get deleted from your online backup account, and if you change a file on your computer, the online version changes, too.

A backup service is great if you want to always keep a huge number of files backed up online. In the event your computer suddenly stops working, you can restore all of those files on a new computer or a different hard drive, and you'll get the same copies you had the last time the backup program stored those files online.

A cloud storage service is less practical as an always-on backup solution and more helpful as a way to back up specific files that you want to have access to from anywhere or share with others. The file versions in the cloud storage account are the same as the versions you uploaded, regardless if you changed them on your computer. Like online backup, you can still download the files again should you need to.

4. How to Choose the Right Cloud Storage Provider

Consider several factors before picking any online cloud backup service.

- Security: Your data must be encrypted to keep it private. If you're concerned about the service itself being able to open your files and see all your backed-up data, go with a service that features "zero-knowledge encryption".
- Price: The cost is determined by how much space you anticipate needing. Many services offer either a trial period or free storage to let you try out their features.
- Compatibility: If you want to access your cloud data from your phone, be sure to pick a cloud storage provider that supports it. Similarly, go with a service that can accept the types of files you want to store online.
- Features: Knowing what features your cloud storage service supports is essential in choosing the right one for you. A comparison of the top free cloud storage services can help you decide between a few of the better ones. Beside that, do some research on the company's websites to see what they offer.
- Ease of use: Uploading and accessing your files on the cloud should be easy. If you want to do this from your desktop, make sure it's simple and won't leave you scratching your head each time you just want to throw some files into your cloud storage account. If it isn't easy to use, look elsewhere.
- Reliability: If a cloud storage service shuts down, you might lose all of your data. Choose a company that would give its users fair warning should they close their doors, or at least offer a way for you to transfer your data elsewhere. Cloud storage services that have been in operation for a long time or that are well known are probably more likely to help out should they decide to shut down the business, but you should read the fine print to see their actual policies.

- Bandwidth: If you're a heavy user, you should also think about bandwidth limitations. Some cloud storage services put a cap on how much data can flow in and/or out of your account on a daily or monthly basis. If you plan to have customers, employees, family or friends download large videos or lots of other files throughout the month, make sure the bandwidth cap isn't prohibitive for you.

New Words

storage	[ˈstɔːrɪdʒ]	*n.*存储
online	[ˌɒnˈlaɪn]	*adj.*在线的，联网的，联机的
extra	[ˈekstrə]	*adj.*额外的，补充的，附加的
stream	[striːm]	*n.*信息流，数据流
		*v.*流，流动
video	[ˈvɪdɪəʊ]	*n.*视频
		*adj.*视频的，影像的
upload	[ˌʌpˈləʊd]	*vt.*上传，上载
retrieve	[rɪˈtriːv]	*vt.*取回；恢复；检索；重新得到
		*n.*恢复，取回；检索
reputable	[ˈrepjʊtəbl]	*adj.*值得尊敬的，声誉好的
document	[ˈdɒkjʊmənt]	*n.*文档
account	[əˈkaʊnt]	*n.*账目，账，账户
website	[ˈwebsaɪt]	*n.*网站
drag-and-drop	[dræg-ænd-drɔp]	*n.*拖曳，（鼠标的）拖放动作
folder	[ˈfəʊldə]	*n.*文件夹
torrent	[ˈtɒrənt]	*n.*种子
procedure	[prəˈsiːdʒə]	*n.*程序，工序，过程
version	[ˈvɜːʃn]	*n.*版本
suddenly	[ˈsʌdənlɪ]	*adv.*意外地，忽然地
copy	[ˈkɒpɪ]	*n.*副本，复制品
		*v.*复制
private	[ˈpraɪvɪt]	*adj.*私有的，私人的，私密的
anticipate	[ænˈtɪsɪpeɪt]	*vt.*预感，预见，预料
comparison	[kəmˈpærɪsn]	*n.*比较，对照
limitation	[ˌlɪmɪˈteɪʃn]	*n.*限制，局限；极限
prohibitive	[prəˈhɪbɪtɪv]	*adj.*禁止的；抑制的；过高的

Phrases

cloud storage	云存储
instead of	（用……）代替……，（是……）而不是……
flash drive	闪存盘；闪盘驱动器
certain of	确信……
log in	登录
desktop application	桌面应用程序
shared link	共享的链接，分享的链接
cloud backup	云备份
zero-knowledge encryption	零知识加密
trial period	试用期
be sure to	必定，务必
shut down	关闭；停下；停工

Text B 参考译文

云 存 储

1. 什么是云存储

云存储是把文件存储在云中（在线）。可以将文件在线保存，而不是将文件保存在本地硬盘、外部硬盘或闪存驱动器上。

使用云存储服务有多种原因。也许是本地硬盘驱动器的磁盘空间不足，在这种情况下，可以将云用作额外的存储。如果想在任何地方播放音乐文件，在家中访问工作文件，轻松共享度假视频等，则可以在线将文件上传到云存储服务。使用云存储的另一个原因是想用设置密码和加密保护重要文件的安全。

简而言之，云存储很有用不仅体现在备份方面，而且体现在以下方面：提高安全性，轻松与他人共享文件或从任意设备（手机、平板电脑或另一台计算机）访问这些文件。

2. 云存储如何工作

将文件上传到互联网并且该文件在那里保存较长时间时，就被认为是云存储。最简单的云存储类型是将某些内容上传到服务器，并可以根据需要再次取回。

信誉良好的云存储服务可保护加密后的文件，并要求输入密码才能访问文件。大多数情况下，云存储账户也可以通过双因素身份验证进行保护，因此，想要访问文件的任何人不仅需要知道密码，还需要知道登录请求后发送给手机的其他代码。

大多数云存储服务允许上传所有类型的文件，包括视频、图片、文档、音乐或其他任何内容。但是，有些有限制，只接受某些类型的文件，例如只接受图像或音乐。云存储服务通常会明示允许和不允许的内容。

不同的云存储服务允许通过不同的方法将文件上传到在线账户。有些仅支持浏览器内上传，这意味着必须登录到云存储服务的网站才能上传数据，但是大多数都具有桌面应用程序，只需将文件拖放到服务的专用文件夹中，这让上传文件更容易。大多数还支持从手机上传图像和视频。

种子云存储服务（在线种子客户端）不那么常见，它不仅可以让你从浏览器下载种子，还可以将文件存储到在线账户中以供日后播放或下载。

将文件在线存储后，根据服务的工作方式，可能会获得以下功能：播放视频和音乐，从移动设备访问文件，通过特殊的共享链接轻松与他人共享文件，下载文件返回到计算机，删除它们以释放账户中的空间，对其进行加密以至于隐藏该服务等。

3. 云存储与云备份

云存储和云备份很容易混淆。它们的工作方式相似且最终结果类似，即文件被在线存储。但是使用这些服务有两个完全不同的原因，了解它们之间的差异很重要，这样才能根据自己的情况进行选择。

云存储是一种选择性备份过程，选择要在线存储的文件，然后将它们发送到在线账户。当删除在计算机上在线备份的文件时，该文件仍位于云存储账户中，因为该文件实际上不再与计算机绑定，它只是在线上传的单个文件。

云备份是在计算机上安装程序并告诉其将特定文件在线备份。备份服务比云存储更进一步，它还将上传对文件所做的任何更改，以便当前版本始终在线存储。换句话说，如果从计算机上删除一个文件，则该文件也可能会从在线备份账户中删除，并且如果在计算机上更改文件，则在线版本也会更改。

如果想始终在线备份大量文件，则备份服务非常有用。万一计算机突然停止工作，可以将所有这些文件还原到新计算机或其他硬盘驱动器上，并且将获得与上一次备份程序将这些文件在线存储时相同的副本。

云存储服务作为永远在线的备份解决方案不太实用，如果想从任何地方访问特定文件或与他人共享文件时，可以用云存储服务来备份这些文件，这是个有用的方法。云存储账户中的文件版本与上传的版本相同，无论是否在计算机上进行了更改。与在线备份一样，仍然可以根据需要再次下载文件。

4. 如何选择合适的云存储提供商

选择任何在线云备份服务之前，请考虑几个因素。

- 安全性：必须对数据进行加密以使其私密。如果担心服务本身能够打开文件并查看所有备份的数据，请使用具有“零知识加密”功能的服务。
- 价格：成本取决于预计需要多少空间。许多服务提供试用期或免费存储，允许试用其功能。
- 兼容性：如果要从手机访问云数据，请确保选择支持该数据的云存储提供商。同样，请使用可以接受要在线存储的文件类型的服务。
- 功能：了解云存储服务支持哪些功能对于选择合适的功能至关重要。比较顶级的免费云存储服务可以帮助你在一些更好的服务之间做出选择。除此之外，在公司的网站上做一些研究，看看它们提供了什么。
- 易于使用：把文件上传到云和在云上访问这些文件应该很容易。如果想在桌面上执行此操作，请确保它很简单并且不会在每次想向云存储账户中扔一些文件时就挠头。如果不容易使用，请找找别的。
- 可靠性：如果云存储服务关闭，则可能会丢失所有数据。选择一家公司，如果他们关闭业务将向用户提供合理的警告或至少提供一种将数据转移到其他地方的方式。已经运行了很长时间或众所周知的云存储服务在决定关闭业务时更有可能提供帮助，但是应该阅读细则以了解其实际策略。
- 带宽：如果你是重量级用户，则还应该考虑带宽限制。某些云存储服务会限制每天或每月可流入和/或流出账户的数据量。如果打算让客户、员工、家人或朋友在整个月中下载大型视频或许多其他文件，请确保带宽上限够用。

Exercises

[Ex. 1] Answer the following questions according to Text A.

1. What does cloud computing allow you to do?
2. What is a private cloud? What does this option do?
3. Where does a community cloud exist? Please give an example.
4. What is a public cloud? How is the storage managed? What does this service offer?
5. How many main types of cloud computing service models are available? What are they commonly known as?
6. What is SaaS?
7. What does IaaS mean? What does this option allow you to do?
8. Why may you be able to reduce your operating costs?

9. What does collaboration in a cloud environment do? If you are working on a project across different locations, what could you use cloud computing to do?
10. What is it important to do before considering cloud computing technology?

[Ex.2] Fill in the following blanks according to Text B.

1. Cloud storage is ________________. Instead of keeping your files on your local hard drive, ____________, or ____________, you can save them online.
2. In short, cloud storage is helpful not only ____________ but also ____________ and the ability to ____________ or access them yourself from anywhere: your phone, tablet, or ____________.
3. The simplest type of cloud storage is ____________ to a server and having the ability to ____________ should you want to.
4. Most cloud storage services let you upload all types of files, ________, pictures, ________, ________, or anything else. However, some are limited to accept only certain kinds of files, such as only ________ or music.
5. Once your files are stored online, depending on how the service works, the features you get might include the ability to ____________, ____________ from your mobile device, easily share the files with others through ____________, download the files back to your computer, __________ to free up space in your account, ____________ so that even the service can not see them, and more.
6. Cloud storage is ____________ where you choose which files to store online, and then you send them to ____________. When you delete a file that you backed up online on your computer, the file is ____________ because it isn't actually tied to your computer anymore, it's just ____________ that you uploaded online.
7. Cloud backup is that you ____________ on your computer and tell it to keep specific files ____________. Going a step further than ____________, a backup service will also ____________ you make to the file so that ____________ is always stored online.
8. A cloud storage service is less practical as ____________ and more helpful as a way to ____________ that you want to have access to from anywhere or ____________.
9. Your data ____________ to keep it private. If you're concerned about the service itself being able to open your files and ____________, go with a service that features ____________.
10. If you're a heavy user, you should also think about ____________. Some cloud storage services ____________ how much data can flow in and/or out of your account ____________.

[Ex. 3] Translate the following terms or phrases from English into Chinese and vice versa.

1. cloud computing ____________ 1. ____________

2. cloud service provider	2. ______
3. hybrid cloud	3. ______
4. private cloud	4. ______
5. cloud backup	5. ______
6. *n.*连通性	6. ______
7. *n.*连续性，继续	7. ______
8. *v.*交互；相互作用	8. ______
9. *vt.*项目，工程，计划	9. ______
10. *n.*设置，安装	10. ______

[Ex. 4] Translate the following sentences into Chinese.

Benefits of Cloud Computing

First of all, "cloud" is just a metaphor to describe the technology. Basically, cloud is nothing but a data center filled with hundreds of components like servers, routers, and storage units. Cloud data centers could be anywhere in the world, also you can access it from anywhere with an Internet-connected device. Why do people use it? Because of the following benefits it has:

- Pay-per-use Model: You only have to pay for the services you use, and nothing more.
- 24/7 Availability: It is always online. There is no such time that you cannot use your cloud service.
- Easily Scalable: It is very easy to scale up and down or turn it off as per customers' needs. For instance, if your website's traffic increases only on Friday nights, you can opt for scaling up your servers that particular day of the week and then scaling down for the rest of the week.
- Security: Cloud computing offers amazing data security. Especially if the data is mission-critical, then that data can be wiped off from local drives and kept on the cloud only for your access to stop it ending up in wrong hands.
- Easily Manageable: You only have to pay subscription fees; all maintenance, up-gradation and delivery of services are completely maintained by the cloud provider. This is backed by the Service-level Agreement (SLA).
- These are the reasons why businesses are adopting cloud computing rather than trying to build their own on-site infrastructure.

[Ex. 5] Fill in the blanks with the words given below.

organizations	focus	distributed	infrastructure	clearest
provider	benefits	accomplish	effective	concept

IaaS, PaaS and SaaS

1. Infrastructure as a Service (IaaS)

IaaS makes ___1___ systems accessible for many organizations by allowing them to host their infrastructure either internally on a private or public cloud. Essentially, they give an organization control over the operating system and to platform that forms the foundation of their software ___2___, but give an external cloud provider control over servers and virtualization technologies that make it possible to deploy that infrastructure.

In the context of a distributed system, this means ___3___ have less to worry about. As you can imagine, without an IaaS, the process of developing and deploying a distributed system becomes much more complex and even costly.

2. Platform as a Service (PaaS)

If IaaS effectively splits responsibilities between the organization and the cloud ___4___ (the "service"), the platform as a Service (PaaS) "outsources" even more to the cloud provider. Essentially, an organization simply has to handle the applications and data, leaving every other aspect of their infrastructure to the platform.

This brings many ___5___, and, in theory, should allow even relatively small engineering teams to take advantage of the benefits of a distributed system. The underlying complexity and heavy lifting that a distributed system brings rests with the cloud provider, allowing an organization's engineers to ___6___on what matters most—shipping code.

3. Software as a Service (SaaS)

SaaS solutions are perhaps the ___7___example of a distributed system. Arguably, given the way we use SaaS today, it's easy to forget that it can be a part of a distributed system. The ___8___ is simple: it's a complete software solution delivered to the end-user.

If you're trying to ___9___ something particularly complex, something which you simply do not have the resources to do yourself, a SaaS solution could be ___10___. Users don't need to worry about installing and maintaining software, they can simply access it via the internet.

Online Resources

二维码	内　　容
	计算机专业常用语法（10）：介词
	在线阅读（1）：Internet of Things （物联网）
	在线阅读（2）：How Does RFID Work? （RFID 如何工作？）

Unit 11

Big Data and Big Data Analytics

Text A

Big Data

扫码听课文

Big data is changing the way people work together within organizations. It is creating a culture in which business and IT leaders must join forces to realize the value from all data. Insights from big data can enable all employees to make better decisions — deepening customer engagement, optimizing operations, preventing threats and fraud, and capitalizing on new sources of revenue.

1. The Big Vs

1.1 Value

This is indeed the holy grail of big data and what we are all looking for. One has to demonstrate value that can be extracted from big or small data in order to justify the investments, whether on big data or on traditional analytics, data warehouse or business intelligence tools, whatever may be the buzzing nomenclature. There seems to be an increasing interest related to the value of big data, as indicated by the number of Google searches looking for similar terms over the last two years.

1.2 Volume

There is no doubt that the information explosion has redefined the connotation of volumes. There are several such staggering statistics going around and it has become increasingly difficult to keep track of the number and magnitude of the prefixes are attached to "bytes" while measuring the volume. Since there is a "helluva lot of data", the term "hellabyte" has been coined beyond petabytes, exabytes, zettabytes and yottabytes. However, since these measures will be superseded by the likes of brontobytes, geopbytes and more, lets move on!

1.3 Velocity

Similarly, velocity refers to the speed at which the data is generated. Some of the factors that exacerbate this trend are the proliferation of social media and the explosion of IoT (Internet of Things). In the context of business operations that have not yet been touched by social media or IoT, the velocity arises from sophisticated enterprise applications that capture each and every minute detail involved in the completion of a particular business process. Enterprise applications have traditionally captured such information but the world has woken up to the power of such information largely in the big data era.

1.4 Variety

The last of the original attributes of big data is variety. Since we are living in an increasingly digital world where technology has invaded into our glasses and watches, the variety of data that is generated is mind-boggling. The computing power available is able to process unstructured text, images, audio, video and data from sensors in the IoT (Internet of Things) world that capture (almost) everything around us. This attribute of big data is more relevant today than it ever was.

1.5 Veracity or Validity

Veracity or validity of data is extremely important and fundamental to the extraction of value from the underlying data. Veracity implies that the data is verifiable and truthful. If this condition is violated, the results can be catastrophic. More importantly, there are several cases in which the data is accurate but may not be valid in the particular context. For instance, if we are trying to ascertain the volume of searches on Google related to big data, we will also obtain results pertaining to the hit single "dangerous" from "big data".

1.6 Visible

Information silos have always existed within enterprises and have been one of the major roadblocks in the attempt to extract value from data. Relevant information should not only exist, but also be visible to the right person at the right time. Actionable data needs to be visible transcending the boundaries of functions, departments and even organizations for value unlocking. Individuals might have believed that information in their hands is power but in the age of big data, collective information available to the world at large is truly omnipotent!

1.7 Visual

We live in an increasingly visual world and the statistics of increase in the number of images and videos shared on the Internet is staggering. According to official statistics, 300 hours of video are uploaded every minute on YouTube. In a business context, appropriate visualization of data is critical for the management to be able to extract value from their limited time, resources and even more limited attention span!

2. More Contenders

In addition to the 7 V's described above, there are several other V's that may be considered.

2.1 Volatility

With more applications such as SnapChat and IoT sensors, we may have data in and out in a snap. Volatility of the underlying data sources may become one of the defining attributes in the future.

2.2 Variability

One of the cornerstones of traditional statistics is standard deviation and variability. Whether or not it makes to an extended list of V's relating to big data, it can never be ignored.

2.3 Viability

Embedded in the concept of value is the need to check the viability of any project. Big data projects can scale up to gigantic proportions and guzzle a lot of resources very quickly. Those who do not learn this fast and get fascinated with fads will funnel funds towards futility resulting in failure. In a nutshell, viability of any project needs to be established and big data projects do not have the liberty of exemption, whether or not it remains a trending buzzword.

2.4 Vitality

Vitality or criticality of the data is another concept that is crucial and is embedded in the concept of value. Information that is more meaningful or critical to the underlying business objective needs to be prioritized. Analysis paralysis needs to be replaced with a more pragmatic approach. Technology allows marketers to create segments of one, but is such extreme segmentation vital or even aligned to the organizational strategy?

2.5 Vincularity

Derived from Latin, it implies connectivity or linkage. This concept is very relevant in today's connected world. There is significant value arbitrage potential by connecting diverse information sets. For instance, the government has forever been trying to connect the details of major expenditure heads and correlating the same with the income declared in tax returns to identify concealment of income. The same purpose may now be achieved by drawing information from social media posts.

3. An Example of Big Data

An example of big data might be petabytes (1024 terabytes) or exabytes (1024 petabytes) of data consisting of billions to trillions of records of millions of people — all from different sources (e.g. web, sales, customer contact center, social media, mobile data and so on). The data is typically loosely structured data that is often incomplete and inaccessible.

New Words

realize	[ˈriːəlaɪz]	*vt.*认识到，了解，实现，实行
engagement	[ɪnˈgeɪdʒmənt]	*n.*参与度，敬业度

fraud	[frɔ:d]	*n.*欺骗,欺诈行为
indeed	[ɪnˈdi:d]	*adv.*真正地,确实；当然
demonstrate	[ˈdemənstreɪt]	*vt.*示范,证明,论证
nomenclature	[nəˈmenklətʃə]	*n.*系统命名法；命名；术语；专门名称
analytics	[ˌænəˈlɪtɪks]	*n.*分析学,解析学,分析论
redefine	[ˌri:dɪˈfaɪn]	*v.*重新定义
staggering	[ˈstægərɪŋ]	*adj.*另人惊愕的，难以置信的
helluva	[ˈheləvə]	*adj.*很大的
hellabyte	[ˈheləbaɪt]	*n.*数据单位，等于 10^{27} 字节
exabyte	['eksəbaɪt]	*n.*数据单位，缩写为 EB，1EB=1024PB
zettabyte	[ˈzetəbaɪt]	*n.*数据单位，缩写为 ZB，1ZB=1024EB
yottabyte	[ˈjɔtəbaɪt]	*n.*数据单位，缩写为 YB，1YB=1024ZB
brontobyte	[ˈbrɔntəbaɪt]	*n.*数据单位，缩写为 BB，1BB=1024YB
geopbyte	[ˈdʒɪəpbaɪt]	*n.*数据单位，缩写为 GB，1GB=1024BB
velocity	[vəˈlɒsətɪ]	*n.*时效性；速度,速率
exacerbate	[ɪgˈzæsəbeɪt]	*vt.*恶化，使加剧
trend	[trend]	*n.*倾向，趋势
proliferation	[prəˌlɪfəˈreɪʃn]	*n.*激增，迅速扩散
explosion	[ɪkˈspləʊʒn]	*n.*爆发，爆炸
era	[ˈɪərə]	*n.*时代，纪元，时期
variety	[vəˈraɪətɪ]	*n.*多样性；品种，种类
mind-boggling	[ˈmaɪndˌbɔglɪŋ]	*adj.*令人难以置信的
unstructured	[ʌnˈstrʌktʃəd]	*adj.*非结构化的，未组织的
veracity	[vəˈræsətɪ]	*n.*真实性
validity	[vəˈlɪdətɪ]	*n.*有效性；合法性,正确性
extremely	[ɪkˈstri:mlɪ]	*adv.*极端地，非常地
fundamental	[ˌfʌndəˈmentl]	*adj.*基础的，基本的 *n.*基本原则，基本原理
verifiable	['verɪfaɪəbl]	*adj.*能证实的
truthful	[ˈtru:θfl]	*adj.*诚实的，说实话的
violate	[ˈvaɪəleɪt]	*vt.*违犯，冒犯，干扰；违反
catastrophic	[ˌkætə'strɒfɪk]	*adj.*悲惨的，灾难的
visible	[ˈvɪzəbl]	*adj.*看得见的，明显的，显著的 *n.*可见物
transcend	[trænˈsend]	*vt.*超越，胜过

boundary	[ˈbaʊndrɪ]	*n.*边界，分界线
omnipotent	[ɒmˈnɪpətənt]	*adj.*全能的，无所不能的
visualization	[ˌvɪʒʊəlaɪ'zeɪʃn]	*n.*可视化
contender	[kənˈtendə]	*n.*竞争者
volatility	[ˌvɒlə'tɪlətɪ]	*n.*波动率；波动性；波动
variability	[ˌveərIəˈbɪlətI]	*n.*变异性；可变性
cornerstone	[ˈkɔ:nəstəʊn]	*n.*奠基石，基础，最重要部分
viability	[ˌvɑɪə'bɪlətɪ]	*n.*可行性，切实可行，能办到；生存能力
gigantic	[dʒaɪˈgæntɪk]	*adj.*巨人般的，巨大的
proportion	[prəˈpɔ:ʃn]	*n.*比例；均衡；部分
		*vt.*使成比例；使均衡，分摊
guzzle	[ˈgʌzl]	*vt.*狂饮，暴食；消耗
fascinate	[ˈfæsɪneɪt]	*vt.*使着迷，使神魂颠倒
		*vi.*入迷,极度迷人的
fad	[fæd]	*n.*时尚，一时流行的狂热，一时的爱好
funnel	[ˈfʌnl]	*vt.&vi.*把……灌进漏斗；使成漏斗状；成漏斗形；使汇集
		*n.*漏斗；漏斗状物
futility	[fju:'tɪlətɪ]	*n.*无益，无用
nutshell	[ˈnʌtʃel]	*n.*简而言之，一言以蔽之
exemption	[ɪgˈzempʃn]	*n.*解除，免除
vitality	[vɑɪˈtælətɪ]	*n.*时效性；动态性，灵活
criticality	[krɪtɪ'kælɪtɪ]	*n.*临界点；临界状态；紧急程度，危险程度
prioritize	[prɑɪˈɒrətaɪz]	*vt.*把……区分优先次序
pragmatic	[prægˈmætɪk]	*adj.*实际的，注重实效的
arbitrage	[ˈɑ:bɪtrɑ:ʒ]	*n.*套汇，套利交易
incomplete	[ˌɪnkəmˈpli:t]	*adj.*不完全的，不完善的

Phrases

big data	大数据
capitalize on	充分利用；资本化
holy grail	圣杯；无处寻觅的稀世珍宝，努力却无法得到的东西
extract from	从……中抽取，从……中提取
data warehouse	数据仓库

business intelligence tool	商业智能工具
information explosion	信息爆炸，知识爆炸
be superseded by	被……取代
wake up	活跃起来；引起注意；（使）认识到
invade into	侵入
unstructured text	非结构化文本
underlying data	源数据；基础数据；基本数据
pertain to	属于，关于，附属
in the attempt to	试图，企图
at large	普遍的；一般的；整体的
according to	依照
in a snap	立刻，马上
standard deviation	标准差，标准偏差
scale up	按比例增加，按比例提高
get fascinated with	迷上，沉溺于
in a nutshell	简而言之，一言以蔽之
analysis paralysis	过度分析
be replaced with	由……代替
be aligned to	与……一致
be derived from	来自，源于
draw from	从……抽取
consist of	构成，组成
customer contact center	客户联络中心，客户服务中心

Abbreviations

IT (Information Technology)	信息技术
IoT (Internet of Things)	物联网

Text A 参考译文

大 数 据

大数据正在改变组织内部人们协同工作的方式。它正在创造一种文化，使得业务和 IT 领导必须联合起来，以便实现所有数据的价值。大数据让所有员工能够做出更好的决策——

深化客户参与度、优化运营、防止威胁和欺诈行为并开辟新的收入来源。

1. 大 V

1.1 价值

这的确是大数据的梦想，也是我们在寻找的目标。从大大小小的数据中获得价值以证明投资所值，无论是大数据分析或传统分析，数据仓库或商业智能工具，或许只是不同的名称而已。根据谷歌搜索过去两年寻找类似条目的数量，似乎表明人们对大数据的价值越来越感兴趣。

1.2 数据量

毫无疑问，信息爆炸已经重新定义数据量的含义。出现了几个非常惊人的统计数字，要跟踪数据越来越难了，要度量这样的数据需要给“字节”前面加上种种前缀。因为有“巨量的数据”，新创造出的术语“hellabyte”已经超越 PB、EB、ZB 和 YB。然而，这些度量单位将被 brontobytes、geopbyte 等替代，让我们继续吧！

1.3 时效性

同样地，时效性是指数据产生的速度。社交媒体的迅速扩散和 IoT（物联网）的爆炸式增长是加剧这一趋势的一些因素。在尚未被社交媒体或物联网影响的业务运营中，时效性来自复杂的企业应用，它捕捉了每一个特定业务流程的每一个微小的细节。传统上企业应用也捕获这些信息，但在大数据时代，这些信息就是力量。

1.4 多样性

最后一个大数据的原始属性是多样性。我们生活在一个日益数字化的世界里，技术已经侵入眼镜和手表，所产生的数据的多样性是令人难以置信的。可用的计算能力能够处理非结构化的文本、图像、音频、视频以及来自 IoT（物联网）传感器的数据，这几乎可以捕获我们周围的一切。如今，大数据的这个属性与现在的生活比以往联系更紧密。

1.5 真实性或有效性

数据的真实性或有效性对提取基础数据的价值非常重要。真实性意味着数据是可验证的和真实的。如果违反这个条件，其结果可能是灾难性的。更重要的是，存在几种情况，其中数据是准确的但在特定情况下无效。例如，如果试图确定谷歌中大数据的搜索量，我们也会获得有关“大数据”的“危险”的结果。

1.6 可见性

信息孤岛一直在企业中存在，并且一直是从数据中提取价值的主要障碍之一。不仅应该有相关信息，而且还应该在合适的时间给合适的人看到。可操作的数据需要超越职能、部门甚至组织的界限是可见的才能释放价值。个人可能会认为在他们手中信息就是力量，但在大数据时代，大量的、对全球有效的整合信息才真正无所不能！

1.7 视觉性

我们生活在一个日益视觉化的世界，统计表明在因特网上共享的图像和视频的数量以惊人的速度增加。据官方统计，每一分钟有 300 小时的视频被上传到 YouTube。在商业环境中，适当的数据可视化对管理者是至关重要的，他们能够在有限的时间、资源甚至更有限的注意力中获得价值！

2. 更多的属性

除了上述的 7 个 V，还有其他几个 V。

2.1 波动率

随着越来越多的应用，如 SnapChat 和物联网传感器，可能即时产生一些输入和输出数据。基础数据源的波动率将来可能成为定义属性之一。

2.2 变异性

传统统计的一个基石是标准差和变异。无论它是否在大数据的 V 属性扩展列表中，都绝不能被忽略。

2.3 可行性

每个项目的可行性都需要检查，这包含在价值概念之中。大数据项目可占据巨大的比例并非常快地消耗大量资源。不快速学习仅沉迷于现状，就会耗尽资金而失败。简而言之，任何项目都要进行可行性研究，大数据项目也不例外，无论它是否仍然是一个流行词。

2.4 时效性

数据的时效性或关键性是另一至关重要的概念，它包含在价值概念之中。应该优先考虑对实现基础商业目标更有意义或更重要的信息。需要用更务实的方法来取代过度分析。技术允许营销人员创建一个片段，但这样极端的分割对组织至关重要吗？它与组织战略一致吗？

2.5 连通性

Vincularity 这个词汇源于拉丁语，意思是连通性或链接。这个概念与当今的互联世界密切相关。连接不同信息集合可以得到潜在的套利价值。例如，政府一直尝试把主要支出的细节相连接，并将其与收入报税单相关联以发现是否隐瞒收入。而现在这一目的可以通过从社交媒体的帖子上提取信息来实现。

3. 一个大数据的示例

大数据的一个例子可能是 PB 级数据（1024 兆兆字节）或艾字节（1024 千兆兆字节），

它包含了数百万人数十亿条的记录——来自不同信息源（如网络、销售、客户联络中心、社交媒体及移动数据等）。该数据通常结构性不强，而且往往是不完整的和难以访问的。

Text B

扫码听课文

Big Data Analytics

Big data analytics is the process of collecting, organizing and analyzing large sets of data to discover patterns and other useful information. Big data analytics can help organizations to better understand the information contained within the data and will also help identify the data that is most important to the business and future business decisions. Analysts working with big data basically want the knowledge that comes from analyzing the data.

1. Big Data Requires High Performance Analytics

To analyze such a large volume of data, big data analytics is typically performed using specialized software tools and applications for predictive analytics, data mining, text mining, forecasting and data optimization. Collectively these processes are separate. Using big data tools and software enables an organization to process extremely large volumes of data that a business has collected to determine which data is relevant and can be analyzed to drive better business decisions in the future.

2. The Challenges of Big Data Analytics

For most organizations, big data analysis is a challenge. Consider the sheer volume of data and the different formats of the data (both structured and unstructured data) that is collected across the entire organization and the many different ways different types of data can be combined, contrasted and analyzed to find patterns and other useful business information.

The first challenge is in breaking down data silos to access all data an organization stores in different places and often in different systems. The second big data challenge is in creating platforms that can pull in unstructured data as easily as structured data. This massive volume of data is typically so large that it's difficult to process using traditional database and software methods.

3. How Big Data Analytics is Used Today

As the technology that helps an organization to break down data silos and analyze data improves, business can be transformed in all sorts of ways. According to Datamation, today's advances in analyzing big data allow researchers to decode human DNA in minutes, predict where terrorists plan to attack, determine which gene is mostly likely to be responsible for certain diseases and, of course, which ads you are most likely to respond to on Facebook.

Another example comes from one of the biggest mobile carriers in the world. France's

Orange launched its Data for Development project by releasing subscriber data for customers in the Ivory Coast. The 2.5 billion records, which were made anonymous, included details on calls and text messages exchanged between 5 million users. Researchers accessed the data and sent Orange proposals for how the data could serve as the foundation for development projects to improve public health and safety. Proposed projects included one that showed how to improve public safety by tracking cell phone data to map where people went after emergencies; another showed how to use cellular data for disease containment.

4. The Benefits of Big Data Analytics

Enterprises are increasingly looking to find actionable insights into their data. Many big data projects originate from the need to answer specific business questions. With the right big data analytics platforms in place, an enterprise can boost sales, increase efficiency, and improve operations, customer service and risk management.

Webopedia parent company, QuinStreet, surveyed 540 enterprise decision-makers involved in big data purchases to learn which business areas companies plan to use big data analytics to improve operations. About half of all respondents said they were applying big data analytics to improve customer retention, help with product development and gain a competitive advantage.

Notably, the business area getting the most attention relates to increasing efficiency and optimizing operations. Specifically, 62 percent of respondents said that they used big data analytics to improve speed and reduce complexity.

5. Top 10 Hot Big Data Technologies

As the big data analytics market rapidly expands, which technologies are most in demand and promise the most growth potential? The answers can be found in TechRadar: Big Data, Q1 2019, a new Forrester Research report evaluating the maturity and trajectory of 22 technologies across the entire data life cycle. The winners all contribute to real-time, predictive, and integrated insights, and this is what big data customers want now.

Here is my talk on the 10 hottest big data technologies based on Forrester's analysis.

5.1 Predictive Analytics

Software and/or hardware solutions that allow firms to discover, evaluate, optimize, and deploy predictive models by analyzing big data sources to improve business performance or mitigate risk.

5.2 NoSQL Databases

Key-value, document, and graph databases.

5.3 Search and Knowledge Discovery

Tools and technologies to support self-service extraction of information and new insights from large repositories of unstructured and structured data that resides in multiple sources such as file systems, databases, streams, APIs, and other platforms and applications.

5.4 Stream Analytics

Software that can filter, aggregate and analyze a high throughput of data from multiple disparate live data sources and in any data format.

5.5 In-memory Data Fabric

Provides low-latency access and processing of large quantities of data by distributing data across the dynamic random access memory (DRAM), Flash, or SSD of a distributed computer system.

5.6 Distributed File Stores

A computer network where data is stored on more than one node, often in a replicated fashion, for redundancy and performance.

5.7 Data Virtualization

A technology that delivers information from various data sources, including big data sources such as Hadoop and distributed data stores in real-time and near-real time.

5.8 Data Integration

Tools for data orchestration across solutions such as Amazon Elastic MapReduce (EMR), Apache Hive, Apache Pig, Apache Spark, MapReduce, Couchbase, Hadoop, and MongoDB.

5.9 Data Preparation

Software that eases the burden of sourcing, shaping, cleansing, and sharing diverse and messy data sets to accelerate data's usefulness for analytics.

5.10 Data Quality

Products that conduct data cleansing and enrichment on large, high-velocity data sets, using parallel operations on distributed data stores and databases.

New Words

analytic	[ˌænəˈlɪtɪk]	*adj.*分析的，解析的
predictive	[prɪˈdɪktɪv]	*adj.*预言性的，成为前兆的
forecasting	['fɔ:kɑ:stɪŋ]	*n.*预测
collectively	[kə'lektɪvlɪ]	*adv.*全体地，共同地
sheer	[ʃɪə]	*adj.*全然的，纯粹的，绝对的
combine	[kəmˈbaɪn]	*vt.*组合，结合
contrast	[ˈkɒntrɑ:st]	*vt.*使与……对比，使与……对照 *vi.*和……形成对照 *n.*对比，对照，(对照中的)差异

silo	[ˈsaɪləʊ]	*n.*竖井
datamation	[ˌdeɪtəˈmeɪʃən]	*n.*自动化资料处理
researcher	[rɪˈsɜːtʃə]	*n.*研究者
predict	[prɪˈdɪkt]	*v.*预知，预言，预报
terrorist	[ˈterərɪst]	*n.*恐怖分子
gene	[dʒiːn]	*n.*[遗传]基因
disease	[dɪˈziːz]	*n.*疾病，弊病
anonymous	[əˈnɒnɪməs]	*adj.*匿名的
emergency	[ɪˈmɜːdʒənsɪ]	*n.*紧急情况，突然事件，非常时刻，紧急事件
containment	[kənˈteɪnmənt]	*n.*控制，遏制政策
retention	[rɪˈtenʃn]	*n.*保持力
mainstream	[ˈmeɪnstriːm]	*n.*主流
maturity	[məˈtʃʊərɪtɪ]	*n.*成热，完备
trajectory	[trəˈdʒektərɪ]	*n.*轨道，弹道
mitigate	[ˈmɪtɪgeɪt]	*v.*减轻
self-service	[ˈselfˈsɜːvɪs]	*n.*自助式
low-latency	[ləʊ-ˈleɪtənsɪ]	*n.*低反应期，短反应时间
orchestration	[ˌɔːkɪˈstreɪʃn]	*n.*管弦乐编曲
burden	[ˈbɜːdn]	*n.*担子，负担 *v.*负担
messy	[ˈmesɪ]	*adj.*凌乱的，杂乱的
usefulness	[ˈjuːsfʊlnɪs]	*n.*有用，有效性

Phrases

high performance	高性能，高精确度
text mining	文本挖掘
in breaking down	在打破
data silo	数据竖井，数据孤岛
subscribe for	预订，认购
text message	短信，短消息
cell phone	手机
disease containment	疾病控制
originate from	发源于
customer service	客户服务
competitive advantage	竞争优势

life cycle	生命周期
business performance	经营成绩，经营业绩
multiple source	多个来源，复合源
distributed computer system	分布式计算机系统
parallel operation	平行工作

Abbreviations

DNA (DeoxyriboNucleic Acid)	脱氧核糖核酸
SSD (Solid State Drives)	固态硬盘

Text B 参考译文

大数据分析

大数据分析是收集、组织和分析大数据集以发现模式和其他有用信息的过程。大数据分析可以帮助组织更好地理解数据中包含的信息，还将帮助辨别哪些数据对于业务和未来业务决策来说最重要。大数据分析师希望获得经分析数据后的知识。

1. 大数据需要高性能分析

为了分析如此大量的数据，通常使用专门的软件工具和应用程序执行大数据分析，以进行预测分析、数据挖掘、文本挖掘、预测和数据优化。总体而言，这些过程是独立的。使用大数据工具和软件可以使组织很好地处理企业收集的大量数据，以确定哪些数据是相关的，并可以进行分析以在将来做出更好的业务决策。

2. 大数据分析的挑战

对于大多数组织而言，大数据分析是一个挑战。想一想整个组织所收集的庞大数据量和不同格式的数据（结构化和非结构化数据），以及为了找出模式和其他有用的业务信息不同类型数据的组合、对比和分析的种种方式。

第一个挑战是打破数据孤岛，以访问组织存储在不同地方和不同系统中的所有数据。第二个大数据的挑战是创建平台，让引入非结构化数据可以像引入结构化数据一样容易。通常，这种海量数据量如此庞大，以至于难以使用传统的数据库和软件方法进行处理。

3. 如今如何使用大数据分析

随着帮助组织打破数据孤岛和分析数据的技术不断改进，可以以各种方式改变业务。根据 Datamation 的研究，当今在分析大数据方面的进步使研究人员可以在几分钟内解码人类DNA，预测恐怖分子计划攻击的地方，确定最有可能导致某些疾病的基因，当然也包括发现你最有可能回应 Facebook 上的哪些广告。

另一个例子来自世界上最大的移动运营商之一。法国的 Orange 通过发布象牙海岸客户的用户订阅数据来启动其“开发数据”项目。这 25 亿条记录是匿名的，其中包括 500 万用户之间通话和短信的详细信息。研究人员访问了这些数据，并向 Orange 发送了有关如何将这些数据用作开发项目的基础以改善公共卫生和安全的建议。拟议的项目包括通过跟踪手机数据以绘制人们在紧急情况下的去向来改善公共安全；另一个展示了如何使用细胞数据来控制疾病。

4. 大数据分析的好处

企业越来越多地寻找对数据的可操作性见解。许多大数据项目源于回答特定业务问题的需求。有了合适的大数据分析平台，企业可以增加销量、提高效率并改善运营、客户服务和风险管理。

Webopedia 的母公司 QuinStreet 对 540 位参与大数据购买的企业决策者进行了调查，以了解公司计划使用大数据分析来改善运营的业务领域。大约一半的受访者表示，他们正在应用大数据分析来提高客户保留率，帮助产品开发并获得竞争优势。

值得注意的是，最受关注的业务领域是提高效率和优化运营。具体来说，有 62%的受访者表示他们使用大数据分析来提高速度和降低复杂性。

5. 十大热门大数据技术

随着大数据分析市场迅速扩展，哪些技术需求最大、增长潜力最大？答案可在《TechRadar：大数据，2019 年第一季度》中找到，该报告是 Forrester Research 的新报告，它评估了整个数据生命周期中 22 种技术的成熟度和发展轨迹。获奖者都为实时、预测性和集成性见解做出了贡献，这正是大数据客户现在想要的。

以下是我根据 Forrester 分析得出的 10 种最热门的大数据技术。

5.1　预测分析

允许公司通过分析大数据源来发现、评估、优化和部署预测模型的软件和/或硬件解决方案，以改善业务绩效或降低风险。

5.2　NoSQL 数据库

键值、文档和图形数据库。

5.3　搜索和知识发现

支持从驻留在多个源（如文件系统、数据库、流、API 和其他平台和应用程序）中的大型非结构化和结构化数据存储库中自助提取信息和新见解的工具和技术。

5.4 流分析

可以过滤、聚合和分析来自多个不同实时数据源的任何数据格式的高吞吐量数据的软件。

5.5 内存中的数据结构

对分布式计算机系统的动态随机存取存储器（DRAM）、闪存或 SSD 中的分布数据，提供低延迟访问和大量数据的处理。

5.6 分布式文件存储

一种计算机网络，其中的数据通常以复制的方式存储在多个节点上，以实现冗余和提高性能。

5.7 数据虚拟化

一种从各种数据源（包括 Hadoop 等大数据源和分布式数据存储）实时和接近实时地传递信息的技术。

5.8 数据整合

跨解决方案的数据编排工具，例如 Amazon Elastic MapReduce（EMR）、Apache Hive、Apache Pig、Apache Spark、MapReduce、Couchbase、Hadoop 和 MongoDB。

5.9 数据准备

该软件可减轻获取、整理、清理和共享各种混乱数据集的负担，从而加快分析时数据的实用性。

5.10 数据质量

通过对分布式数据存储和数据库并行操作，对大型、高速数据集进行数据清理和浓缩的产品。

Exercises

[Ex. 1] Answer the following questions according to Text A.

1. What can insights from big data do?
2. What does velocity refer to? What are some of the factors that exacerbate this trend?
3. Why is the variety of data that is generated mind-boggling?
4. What does veracity imply?
5. What have always existed within enterprises and have been one of the major roadblocks in the attempt to extract value from data?
6. What should relevant information be?
7. How many hours of video are uploaded every minute on YouTube according to official

statistics?

8. What is one of the cornerstones of traditional statistics?

9. What kind of information needs to be prioritized?

10. Where is the word vincularity derived from? What does it imply?

[Ex. 2] Answer the following questions according to Text B.

1. What is big data analytics?
2. What can big data analytics do?
3. How is big data analytics typically performed to analyze such a large volume of data?
4. What is the first challenge of big data analytics?
5. What is a second big data challenge?
6. What do today's advances in analyzing big data allow researchers to do according to Datamation?
7. What do many big data projects originate from?
8. What can an enterprise do with the right big data analytics platforms in place?
9. What does the business area getting the most attention relate to?
10. What does the last part of the passage mainly talk about?

[Ex. 3] Translate the following terms or phrases from English into Chinese and vice versa.

1. business intelligence tool	1. ______
2. data warehouse	2. ______
3. invade into	3. ______
4. unstructured text	4. ______
5. distributed computer system	5. ______
6. *n.*分析学，解析学，分析论	6. ______
7. *adj.*基础的，基本的	7. ______
8. *n.*欺骗，欺诈行为	8. ______
9. *v.*重新定义	9. ______
10. *n.*有效性；合法性，正确性	10. ______

[Ex. 4] Translate the following sentences into Chinese.

Big Data Analytics

Big data analytics is the use of advanced analytic techniques against very large, diverse data sets that include structured, semi-structured and unstructured data, from different sources, and in different sizes.

Big data is a term applied to data sets whose size or type is beyond the ability of traditional relational databases to capture, manage and process the data with low latency. Big data has one or more of the following characteristics: high volume, high velocity or high variety. Artificial intelligence (AI), mobile, social and the Internet of Things (IoT) are driving data complexity through new forms and sources of data. For example, big data comes from sensors, devices, video/audio, networks, log files, transactional applications, web, and social media — much of it generated in real time and at a very large scale.

Analysis of big data allows analysts, researchers and business users to make better and faster decisions using data that was previously inaccessible or unusable. Businesses can use advanced analytics techniques such as text analytics, machine learning, predictive analytics, data mining, statistics and natural language processing to gain new insights from previously untapped data sources independently or together with existing enterprise data.

[Ex. 5] Fill in the blanks with the words given below.

relational	structured	considerably	unstructured	market
columns	sorted	field	numbers	compared

Three Different Data Structures

1. Structured Data

Structured data is data that adheres to a pre-defined data model and is therefore straightforward to analyse. Structured data conforms to a tabular format with relationship between the different rows and ___1___. Common examples of structured data are Excel files or SQL databases. Each of these have structured rows and columns that can be ___2___.

Structured data depends on the existence of a data model—a model of how data can be stored, processed and accessed. Because of a data model, each ___3___ is discrete and can be accesses separately or jointly along with data from other fields. This makes structured data extremely powerful: it is possible to quickly aggregate data from various locations in the database.

Structured data is is considered the most "traditional" form of data storage, since the earliest versions of database management systems (DBMS) were able to store, process and access ___4___ data.

2. Unstructured Data

Unstructured data is information that does not have a pre-defined data model is not organised in a pre-defined manner. Unstructured information is typically text-heavy, but may contain data such as dates, ___5___ numbers, and facts as well. This results in irregularities and ambiguities that make it difficult to understand using traditional programs as ___6___ to data stored in structured databases. Common examples of unstructured data include audio, video files or No-SQL databases.

The ability to store and process unstructured data has greatly grown in recent years, with many new technologies and tools coming to the ____7____ that are able to store specialised types of unstructured data. MongoDB, for example, is optimised to store documents. Apache Giraph, as an opposite example, is optimised for storing relationships between nodes.

The ability to analyse unstructured data is especially relevant in the context of big data, since a large part of data in organisations is ____8____. Think about pictures, videos or PDF documents. The ability to extract value from unstructured data is one of main drivers behind the quick growth of big data.

3. Semi-structured Data

Semi-structured data is a form of structured data that does not conform with the formal structure of data models associated with ____9____ databases or other forms of data tables, but nonetheless contain tags or other markers to separate semantic elements and enforce hierarchies of records and fields within the data. Therefore, it is also known as self-describing structure. Examples of semi-structured data include JSON and XML are forms of semi-structured data.

The reason that this third category exists (between structured and unstructured data) is because semi-structured data is ____10____ easier to analyse than unstructured data. Many big data solutions and tools have the ability to "read" and process either JSON or XML. This reduces the complexity to analyse structured data, compared to unstructured data.

Online Resources

二维码	内　　容
	计算机专业常用语法（11）：数词
	在线阅读（1）：Structured Data,Semi-structured Data,Unstructured Data （结构化数据、半结构化数据、非结构化数据）
	在线阅读（2）：Big Data Security （大数据安全）

Unit 12

Artificial Intelligence

Text A

Artificial Intelligence (AI)

Artificial intelligence (AI) is the simulation of human intelligence processes by machines, especially computer systems. These processes include learning (the acquisition of information and rules for using the information), reasoning (using rules to reach approximate or definite conclusions) and self-correction. Particular applications of AI include expert systems, speech recognition and machine vision.

AI can be categorized as either weak or strong. Weak AI, also known as narrow AI, is an AI system that is designed and trained for a particular task. Virtual personal assistants, such as Apple's Siri, are a form of weak AI. Strong AI, also known as artificial general intelligence, is an AI system with generalized human cognitive abilities. When presented with an unfamiliar task, a strong AI system is able to find a solution without human intervention.

Because hardware, software and staffing costs for AI can be expensive, many vendors are including AI components in their standard offerings, as well as access to artificial intelligence as a service (AIaaS) platforms. AI as a service allows individuals and companies to experiment with AI for various business purposes and sample multiple platforms before making a commitment. Popular AI cloud offerings include Amazon AI services, IBM Watson Assistant, Microsoft Cognitive Services and Google AI services.

While AI tools present a range of new functionality for businesses, the use of artificial intelligence raises ethical questions. This is because deep learning algorithms, which underpin many of the most advanced AI tools, are only as smart as the data they are given in training. Because a human selects what data should be used for training an AI program, the potential for human bias is inherent and must be monitored closely.

Some industry experts believe that the term artificial intelligence is too closely linked to popular culture, causing the general public to have unrealistic fears about artificial intelligence and improbable expectations about how it will change the workplace and life in general. Researchers and marketers hope augmented intelligence, which has a more neutral connotation, will help people understand that AI will simply improve products and services and will not replace the humans that use them.

1. Types of AI

Arend Hintze, an assistant professor of integrative biology and computer science and engineering at Michigan State University, categorizes AI into four types, from the kind of AI systems that exist today to sentient systems which do not yet exist. His categories are as follows.

- Type 1: Reactive machines. An example is Deep Blue, the IBM chess program that beat Garry Kasparov in the 1990s. Deep Blue can identify pieces on the chess board and make predictions, but it has no memory and cannot use past experiences to inform future ones. It analyzes possible moves—its own and its opponent—and chooses the most strategic move. Deep Blue and Google's AlphaGo were designed for narrow purposes and cannot easily be applied to another situation.
- Type 2: Limited memory. These AI systems can use past experiences to inform future decisions. Some of the decision-making functions in self-driving cars are designed this way. Observations inform actions happening in the not-so-distant future, such as a car changing lanes. These observations are not stored permanently.
- Type 3: Theory of mind. This psychology term refers to the understanding that others have their own beliefs, desires and intentions that impact the decisions they make. This kind of AI does not yet exist.
- Type 4: Self-awareness. In this category, AI systems have a sense of self, have consciousness. Machines with self-awareness understand their current state and can use the information to infer what others are feeling. This type of AI does not yet exist.

2. Examples of AI Technology

AI is incorporated into a variety of different types of technology. Here are some examples.

(1) Automation: What makes a system or process function automatically. For example, robotic process automation (RPA) can be programmed to perform high-volume, repeatable tasks that humans normally performed. RPA is different from IT automation in that it can adapt to changing circumstances.

(2) Machine learning: The science of getting a computer to act without programming. Deep learning is a subset of machine learning that, in very simple terms, can be thought of as the automation of predictive analytics. There are three types of machine learning algorithms:

- Supervised learning: Data sets are labeled so that patterns can be detected and used to label new data sets.
- Unsupervised learning: Data sets aren't labeled and are sorted according to similarities or differences.
- Reinforcement learning: Data sets aren't labeled but, after performing an action or several actions, the AI system is given feedback.

(3) Machine vision: The science of allowing computers to see. This technology captures

and analyzes visual information using a camera, analog-to-digital conversion and digital signal processing. It is often compared to human eyesight, but machine vision isn't bound by biology and can be programmed to see through walls, for example. It is used in a range of applications from signature identification to medical image analysis. Computer vision, which is focused on machine-based image processing, is often conflated with machine vision.

(4) Natural language processing (NLP): The processing of human—and not computer—language by a computer program. One of the older and best known examples of NLP is spam detection, which looks at the subject line and the text of an email and decides if it's junk. Current approaches to NLP are based on machine learning. NLP tasks include text translation, sentiment analysis and speech recognition.

(5) Robotics: A field of engineering focused on the design and manufacturing of robots. Robots are often used to perform tasks that are difficult for humans to perform or perform consistently. They are used in assembly lines for car production or by NASA to move large objects in space. Researchers are also using machine learning to build robots that can interact in social settings.

(6) Self-driving cars: These use a combination of computer vision, image recognition and deep learning to build automated skill at piloting a vehicle while staying in a given lane and avoiding unexpected obstructions, such as pedestrians.

3. AI Applications

AI has made its way into a number of areas. Here are six examples.

- AI in healthcare. The biggest bets are on improving patient outcomes and reducing costs. Companies are applying machine learning to make better and faster diagnoses than humans. One of the best known healthcare technologies is IBM Watson. It understands natural language and is capable of responding to questions asked of it. The system mines patient data and other available data sources to form a hypothesis, which then presents with a confidence scoring schema. Other AI applications include chatbots, a computer program used online to answer questions and assist customers, to help schedule follow-up appointments or aid patients through the billing process, and virtual health assistants that provide basic medical feedback.
- AI in business. Robotic process automation is being applied to highly repetitive tasks normally performed by humans. Machine learning algorithms are being integrated into analytics and CRM platforms to uncover information on how to better serve customers. Chatbots have been incorporated into websites to provide immediate service to customers. Automation of job positions has also become a talking point among academics and IT analysts.
- AI in education. AI can automate grading, giving educators more time. AI can assess students and adapt to their needs, helping them work at their own pace. AI tutors can provide additional support to students, ensuring they stay on track. AI could change

where and how students learn, perhaps even replacing some teachers.

- AI in finance. AI in personal finance applications, such as Mint or Turbo Tax, is making a break in financial institutions. Applications such as these collect personal data and provide financial advice. Other programs, such as IBM Watson, have been applied to the process of buying a home. Today, software performs much of the trading on Wall Street.
- AI in law. The discovery process, sifting through documents, in law is often overwhelming for humans. Automating this process is a more efficient use of time. Startups are also building question-and-answer computer assistants that can sift programmed-to-answer questions by examining the taxonomy and ontology associated with a database.
- AI in manufacturing. This is an area that has been at the forefront of incorporating robots into the workflow. Industrial robots used to perform single tasks and were separated from human workers, but as the technology advanced that changed.

4. Security and Ethical Concerns

The application of AI in the realm of self-driving cars raises security as well as ethical concerns. Cars can be hacked, and when an autonomous vehicle is involved in an accident, liability is unclear. Autonomous vehicles may also be put in a position where an accident is unavoidable, forcing the programming to make an ethical decision about how to minimize damage.

Another major concern is the potential for abuse of AI tools. Hackers are starting to use sophisticated machine learning tools to gain access to sensitive systems, complicating the issue of security beyond its current state.

Deep learning-based video and audio generation tools also present bad actors with the tools necessary to create so-called deepfakes, convincingly fabricated videos of public figures saying or doing things that never took place.

5. Regulation of AI Technology

Despite these potential risks, there are few regulations governing the use of AI tools, and where laws do exist, they typically pertain to AI only indirectly. For example, Federal Fair Lending Regulations require financial institutions to explain credit decisions to potential customers, which limit the extent to which lenders can use deep learning algorithms, which by their nature are typically opaque. Europe's GDPR puts strict limits on how enterprises can use consumer data, which impedes the training and functionality of many consumer-orientated AI applications.

In 2016, the National Science and Technology Council (NSTC) issued a report examining the potential role governmental regulation might play in AI development, but it did not recommend specific legislation be considered. Since that time the issue has received little attention from lawmakers.

New Words

simulation	[ˌsɪmjʊˈleɪʃn]	*n.*模仿，模拟
acquisition	[ˌækwɪˈzɪʃn]	*n.*获得
rule	[ru:l]	*n.*规则，规定；统治，支配
		*v.*控制，支配
reasoning	[ˈri:zənɪŋ]	*n.*推理，论证
		*v.*推理，思考；争辩；说服
		*adj.*推理的
approximate	[əˈprɒksɪmɪt]	*adj.*极相似的
		*vi.*接近于，近似于
		*vt.*靠近，使接近
definite	[ˈdefɪnɪt]	*adj.*明确的；一定的；肯定
conclusion	[kənˈklu:ʒn]	*n.*结论；断定，决定；推论
self-correction	[ˌselfkə'rekʃn]	*n.*自校正；自我纠错；自我改正
particular	[pəˈtɪkjʊlə]	*adj.*特别的；详细的；独有的
		*n.*特色，特点
vision	[ˈvɪʒn]	*n.*视觉
narrow	[ˈnærəʊ]	*adj.*狭隘的，狭窄的
virtual	[ˈvɜ:tʃuəl]	*adj.*（计算机）虚拟的；实质上的，事实上的
cognitive	[ˈkɒgnɪtɪv]	*adj.*认知的，认识的
unfamiliar	[ˌʌnfəˈmɪlɪə]	*adj.*不熟悉的；不常见的；陌生的；没有经验的
experiment	[ɪkˈsperɪmənt]	*n.*实验，试验；尝试
		*vi.*做实验
commitment	[kəˈmɪtmənt]	*n.*承诺，许诺；委任，委托
ethical	[ˈeθɪkl]	*adj.*道德的，伦理的
underpin	[ˌʌndəˈpɪn]	*vt.*加固，支撑
unrealistic	[ˌʌnrɪəˈlɪstɪk]	*adj.*不切实际的；不现实的；空想的
fear	[fɪə]	*n.*害怕；可能性
		*vt.*害怕；为……忧虑（或担心、焦虑）
		*vi.*害怕；忧虑
expectation	[ˌekspekˈteɪʃn]	*n.*期待；预期
neutral	[ˈnju:trəl]	*adj.*中立的
connotation	[ˌkɒnəˈteɪʃn]	*n.*内涵，含义
integrative	['ɪntɪgreɪtɪv]	*adj.*综合的，一体化的

sentient	[ˈsentɪənt]	*adj.*有感觉能力的，有知觉力的
reactive	[rɪˈæktɪv]	*adj.*反应的
prediction	[prɪˈdɪkʃn]	*n.*预测，预报；预言
opponent	[əˈpəʊnənt]	*n.*对手
observation	[ˌɒbzəˈveɪʃn]	*n.*观察，观察力
psychology	[saɪˈkɒlədʒɪ]	*n.*心理学；心理特点；心理状态
intention	[ɪnˈtenʃn]	*n.*意图，目的；意向
self-awareness	[self-əˈweənɪs]	*n.*自我意识
consciousness	[ˈkɒnʃəsnɪs]	*n.*意识，观念；知觉
circumstance	[ˈsɜːkəmstəns]	*n.*环境，境遇
similarity	[ˌsɪmɪˈlærɪtɪ]	*n.*相像性，相仿性，类似性
signature	[ˈsɪgnɪtʃə]	*n.*签名；署名；识别标志
identification	[aɪˌdentɪfɪˈkeɪʃn]	*n.*鉴定，识别
detection	[dɪˈtekʃn]	*n.*检查，检测
junk	[dʒʌŋk]	*vt.*丢弃，废弃
		*n.*废品；假货
consistently	[kən'sɪstəntlɪ]	*adv.*一贯地，坚持地
pilot	[ˈpaɪlət]	*n.*引航员；向导
		*vt.*驾驶
vehicle	[ˈviːɪkl]	*n.*车辆；交通工具
unexpected	[ˌʌnɪkˈspektɪd]	*adj.*意外的；忽然的；突然的
obstruction	[əbˈstrʌkʃn]	*n.*阻塞，阻碍，受阻
pedestrian	[pəˈdestrɪən]	*n.*步行者，行人
		*adj.*徒步的
healthcare	['helθkeə]	*n.*卫生保健
diagnose	[ˈdaɪəgnəʊz]	*vt.*诊断；判断
		*vi.*做出诊断
hypothesis	[haɪˈpɒθəsɪs]	*n.*假设，假说；前提
chatbot	[tʃætbɒt]	*n.*聊天机器人
appointment	[əˈpɔɪntmənt]	*n.*预约
repetitive	[rɪˈpetɪtɪv]	*adj.*重复的，啰嗦的
overwhelming	[ˌəʊvəˈwelmɪŋ]	*adj.*势不可挡的，压倒一切的
taxonomy	[tækˈsɒnəmɪ]	*n.*分类学，分类系统
ontology	[ɒnˈtɒlədʒɪ]	*n.*本体，存在；实体论

forefront	[ˈfɔ:frʌnt]	*n.*前列；第一线；活动中心
incorporating	[ɪnˈkɔ:pəreɪtɪŋ]	*v.*融合，包含；使混合
realm	[relm]	*n.*领域，范围
unclear	[ˌʌnˈklɪə]	*adj.*不清楚的，不明白的，含糊不清
unavoidable	[ˌʌnəˈvɔɪdəbl]	*adj.*不可避免的，不得已的
minimize	[ˈmɪnɪmaɪz]	*vt.*把……减至最低数量[程度]，最小化
damage	[ˈdæmɪdʒ]	*n.*损害，损毁；赔偿金
		*v.*损害，毁坏
deepfake	[ˈdi:pfeɪk]	*n.*换脸术
convincingly	[kən'vɪnsɪŋlɪ]	*adv.*令人信服地，有说服力地
fabricate	[ˈfæbrɪkeɪt]	*vt.*编造，捏造
regulation	[ˌregjʊˈleɪʃn]	*n.*规章，规则
		*adj.*规定的
credit	[ˈkredɪt]	*n.*信誉，信用；[金融]贷款
		*vt.*相信，信任
opaque	[əʊˈpeɪk]	*adj.*不透明的；含糊的
		*n.*不透明
impede	[ɪmˈpi:d]	*vt.*阻碍；妨碍；阻止
lawmaker	[ˈlɔ:meɪkə]	*n.*立法者

Phrases

human intelligence	人类智能
expert system	专家系统
speech recognition	语音识别
machine vision	机器视觉
weak AI	弱人工智能
virtual personal assistant	虚拟个人助理
strong AI	强人工智能
artificial general intelligence	通用人工智能
for ... purpose	为了……目的
a range of	一系列，一些，一套
deep learning algorithm	深度学习算法
computer science	计算机科学
sentient system	感觉系统

self-driving car	自动驾驶汽车
not-so-distant future	不远的将来
a sense of ...	一种……感觉
be incorporated into ...	被并入……
predictive analytic	预测分析
supervised learning	有监督学习
unsupervised learning	无监督学习
reinforcement learning	强化学习
analog-to-digital conversion	模（拟）-数（字）转换
digital signal	数字信号
medical image analysis	医学图像分析
machine-based image processing	基于机器的图像处理
be conflated with ...	与……混为一谈
spam detection	垃圾邮件检测
text translation	文本翻译
sentiment analysis	情感分析，倾向性分析
assembly line	（工厂产品的）装配线，流水线
social setting	社会环境，社会场景，社会情境
image recognition	图像识别
confidence scoring schema	置信评分模式
virtual health assistant	虚拟健康助理
talking point	话题；论题；论据
financial institution	金融机构

Abbreviations

AI (Artificial Intelligence)	人工智能
AIaaS (Artificial Intelligence as a Service)	人工智能即服务
RPA (Robotic Process Automation)	机器人流程自动化
NLP (Natural Language Processing)	自然语言处理
NASA (National Aeronautics and Space Administration)	美国航空航天局
CRM (Customer Relationship Management)	客户关系管理
GDPR (General Data Protection Regulation)	普通数据保护条例
NSTC (National Science and Technology Council)	国家科学技术委员会

Text A 参考译文

人 工 智 能

人工智能（AI）是机器，特别是计算机系统对人类智能处理的模拟。这些过程包括学习（获取信息和使用信息的规则）、推理（使用规则来达到近似或明确的结论）和自我校正。人工智能的典型应用包括专家系统、语音识别和机器视觉。

人工智能可以分为弱人工智能与强人工智能两类。弱人工智能，也称为窄人工智能，是为特定任务而设计和训练的人工智能系统。虚拟个人助理，如 Apple 的 Siri，是一种弱人工智能。强人工智能，也称为通用人工智能，是一种具有广泛人类认知能力的人工智能系统。当提出一项不熟悉的任务时，强人工智能系统能够在没有人为干预的情况下找到解决方案。

由于人工智能的硬件、软件和人员成本可能很昂贵，因此许多供应商在其标准产品中包含人工智能组件以及访问人工智能即服务（AIaaS）平台。在做出承诺之前，人工智能即服务允许个人和公司为各种商业目的进行人工智能试验，并对多个平台进行抽样调查。流行的人工智能云产品包括 Amazon AI 服务、IBM Watson Assistant、Microsoft Cognitive Services 和 Google AI 服务。

虽然人工智能工具为企业提供了一系列新功能，但人工智能的使用引发了伦理问题。这是因为深度学习算法是许多最先进的人工智能工具的基础，它们的智能仅仅与训练时所提供的数据匹配。因为由人类选择用何种数据来训练人工智能程序，而人类本身可能有偏见，所以必须密切监控。

一些业内专家认为，人工智能这一术语与流行文化联系太紧密，导致普通大众对人工智能产生不切实际的恐惧，以及对人工智能如何改变工作场所和生活方式所抱有不太可能的期望。研究人员和营销人员希望增强智能（具有更中性内涵）会帮助人们明白人工智能只能改进产品和服务，而不是取代使用它们的人。

1. 人工智能的类型

密歇根州立大学综合生物学和计算机科学与工程的助理教授 Arend Hintze 将人工智能分为 4 类，从现有的人工智能系统到尚未存在的感觉系统。他的分类如下。

- 类型 1：反应机器。一个例子是 Deep Blue（深蓝），它是一个在 20 世纪 90 年代击败 Garry Kasparov 的 IBM 国际象棋程序。Deep Blue 可以识别棋盘上的棋子并进行预测，但它没有记忆，也无法使用过去的经验来指导未来的棋子。它分析了自己和对手可能走的棋，并选择最具战略性的举措。Deep Blue 和 Google 的 AlphaGo 专为狭窄目的而设计，不能轻易应用于其他情况。

- 类型 2：有限的存储。这些人工智能系统可以使用过去的经验来指导未来的决策。自动驾驶汽车的一些决策功能就是这样设计的。观察结果可以告知在不远的将来发生的行动，例如换车道。这些观察结果不会被永久存储。
- 类型 3：心智理论。这个心理学术语指的是他人有自己的信念、欲望和意图，这会影响他们的决策。这种人工智能尚不存在。
- 类型 4：自我意识。在这个类别中，人工智能系统具有自我意识感和知觉。具有自我意识的机器了解其当前状态，并可以使用该信息来推断其他人的感受。这种类型的人工智能尚不存在。

2. 人工智能技术的例子

人工智能被整合到各种不同类型的技术中。这里有一些例子。

（1）自动化：可以使系统或过程自动运行。例如，机器人过程自动化（RPA）可以通过编程来执行人类通常执行的大量、可重复的任务。RPA 与 IT 自动化的不同之处在于它可以适应不断变化的环境。

（2）机器学习：使计算机无须编程即可行动的科学。深度学习是机器学习的一个子集，简而言之，它可以被认为是自动化进行预测分析。有三种类型的机器学习算法。

- 监督学习：标记数据集，以便可以检测模式并用于标记新数据集。
- 无监督学习：不标记数据集，并根据相似性或差异性进行排序。
- 强化学习：不标记数据集，但在执行一个行动或多个行动后，人工智能系统会得到反馈。

（3）机器视觉：让计算机具有视觉的科学。该技术使用相机、模数转换和数字信号处理来捕获和分析视觉信息。它通常被比作人类的视力，但机器视觉不受生物学的约束，例如可以编程以透视墙壁。它用于从签名识别到医学图像分析的各种应用中。计算机视觉是基于机器的图像处理，通常与机器视觉混为一谈。

（4）自然语言处理（NLP）：通过计算机程序处理人类（而不是计算机）的语言。其中一个较早且最著名的 NLP 示例是垃圾邮件检测，它查看主题行和电子邮件的文本并确定它是否为垃圾邮件。目前的 NLP 方法基于机器学习。NLP 任务包括文本翻译、情感分析和语音识别。

（5）机器人技术：一个专注于机器人设计和制造的工程领域。机器人通常用于执行人类难以执行或一直执行的任务。它们用于汽车生产的装配线或由 NASA 用于在太空中移动大型物体。研究人员还利用机器学习来构建可在社交场合进行交互的机器人。

（6）自动驾驶汽车：它们把计算机视觉、图像识别和深度学习相结合，使用自动化技能驾驶车辆，遇到意外障碍（例如行人）时在给定车道上停车。

3. 人工智能应用

人工智能已经进入了许多领域。这里列举 6 个示例。

- 人工智能应用于医疗保健领域。最大的好处是改善患者的治疗效果和降低成本。公司正在应用机器学习来做出比人类更好、更快的诊断。IBM Watson 是最著名的医疗保健技术之一。它理解自然语言，并能够回答所提出的问题。系统挖掘患者数据和其他可用数据源以形成假设，然后它将给出一个置信评分模式。其他人工智能应用程序包括聊天机器人。聊天机器人是一个计算机程序，用于在线回答问题和帮助客户，帮助安排后续预约或自动计费，以及提供基本医疗反馈的虚拟健康助理。
- 人工智能应用于商业领域。机器人过程自动化正被应用于通常由人类执行的、高度重复的任务。机器学习算法正在集成到分析和客户关系管理平台中，用于发现和分析信息并更好地为客户服务。聊天机器人已被纳入网站为客户提供即时服务。工作岗位的自动化也成为学术界和 IT 分析师的话题。
- 人工智能应用于教育领域。人工智能可以自动评分，节省教师时间。人工智能可以评估学生并应对他们的需求，帮助他们按照自己的进度工作。人工智能导师可以为学生提供额外的支持，确保他们处于正确轨道上。人工智能可以改变学生学习的地点和方式，甚至可以取代一些教师。
- 人工智能应用于金融领域。个人理财应用程序中的人工智能（如 Mint 或 Turbo Tax）正在进入金融机构。这些应用程序收集个人数据并提供财务建议。其他程序（例如 IBM Watson）已经应用于购买房屋的过程。如今，华尔街很大一部分交易都是由软件完成的。
- 人工智能应用于法律领域。在法律上，对人来说，发现过程（筛选文件）是非常困难的。自动化地完成此项工作可以大大节省时间。创业公司还在构建计算机回答助手，通过编程来检查与数据库相关的分类和本体，筛选出问题的答案。
- 人工智能应用于制造业。这个领域一直处于将机器人纳入工作流程的最前沿。工业机器人曾经执行单一任务并与人类工作人员分开，但随着技术的进步这一现象已经发生了变化。

4. 安全和伦理问题

人工智能在自动驾驶汽车领域的应用带来了安全和伦理方面的问题。汽车可以被黑客入侵，当自动驾驶汽车涉及事故时，责任并不清楚。自动驾驶车辆也可能处于无法避免事故的情况，迫使编程人员就如何最大限度地减少损坏做出伦理决定。

另一个主要问题是存在滥用人工智能工具的可能性。黑客们开始使用复杂的机器学习工具来访问敏感系统，使安全问题越来越复杂化。

基于深度学习的视频和音频生成工具也为不良行为者提供了所谓换脸所需的工具，他们可以制作公众人物的视频，尽管这些公众人物从未说过这些话，也从未做过这些事，但这些视频却让人不得不信。

5. 人工智能技术的规范

尽管存在这些潜在的风险，但很少有关于人工智能工具使用的法规，而且即便有法规，它们通常也只是间接地涉及人工智能。例如，联邦公平贷款法规要求金融机构向潜在客户解释信用决策，这些法规限制了贷方可以使用深度学习算法的程度，这些算法本质上通常是不透明的。欧洲的 GDPR 严格限制企业使用消费者数据的方法，这阻碍了许多面向消费者的人工智能应用程序的培训和功能。

2016 年，国家科学技术委员会发布了一份报告，研究政府监管在人工智能发展中可能发挥的作用，但并未建议考虑具体立法。从那时起，这个问题就很少受到立法者的关注。

Text B

What Is Machine Learning

扫码听课文

Artificial intelligence and machine learning are among the most trending technologies these days. Artificial intelligence teaches computers to behave like a human, to think, and to give a response like a human, and to perform the actions like humans perform.

1. What Is Machine Learning

As the name suggests, machine learning means the machine is learning.

This is the technique through which we teach the machines about things. It is a branch of artificial intelligence and I would say it is the foundation of artificial intelligence. Here we train our machines using data. If you take a look into it, you'll see that it is something like data mining. Actually, the concept behind it is that machine learning and data mining are both data-oriented. We work on data in both of situations. Actually, in data sciences or big data, we analyze the data and make the statistics out of it and we work on how we can maintain our data, how we can conclude the results and make a summary of it instead of maintaining the complete comprehensive bulk of data. But in machine learning, we teach the machines to make the decisions about things. We teach the machine with different data sets and then we check the machine for some situations and see what kind of results we get from this unknown scenario. We also use this trained model for prediction in new scenarios.

We teach the machine with our historical data, observations, and experiments. And then, we predict with the machine from these learnings and take the responses.

As I already said, machine learning is closely related to data mining and statistics.

- Data mining: Concerned with analytics of data.
- Statistics: Concerned with prediction-making/probability.

2. Why Do We Need Machine Learning

In this era, we're using wireless communication, internet etc. Using social media, or driving cars, or anything we're doing right now, is actually generating the data at the backend. If you're surprised about how our cars are generating the data, remember that every car has a small computer inside which controls your vehicle completely, i.e., when which component needs the current, when the specific component needs to start or switch. In this way, we're generating TBs (terabytes) of data.

But this data is also important to get to the results. Let's take an example and try to understand the concept clearly. Let's suppose a person is living in a town and he goes to a shopping mall and buys something. We have many items of a single product. When he buys something, now we can generate the pattern of the things he has bought. In the same way, we can generate the selling and purchasing patterns of things of different people. Now you might be thinking about a random person who comes and buys something and then he never comes again, but we have the pattern of things there as well. With the help of this pattern we can make a decision about the things people most like and when they come to the mall again. They will see the things they want just at the entrance. This is how we attract the customer with machine learning.

3. How Do Machines Learn

Actually, machines learn through the patterns of data. Let's start with the data sets of data. The input we give to the machine is called X and the response we get is Y. Here we've three types of learning.

- Supervised Learning
- Unsupervised Learning
- Reinforcement Learning

3.1 Supervised Learning

In supervised learning, we know about the different cases (inputs) and we know the labels (output) of these cases. And here we already know about the basic truths, so here we just focus on the function (operation) because it is the main and most important thing here (see Figure 12-1).

Here we just create the function to get the output of the inputs. And we try to create the function which processes the data and try to give the accurate outputs (Y) in most of the scenarios (see Figure 12-2).

Because we've started with known values for our inputs, we can validate the model and make it even better.

And now we teach our machine with different data sets. Now it is the time to check it in unknown cases and generate the value.

Figure 12-1 Inputs and output of supervised learning

Figure 12-2 Function of supervised learning

Note: Let's suppose you've provided the machine a data set of some kind of data and now you train the model according to this data set. Now the result comes to you from this model on the basis of this knowledge set you've provided. But let's suppose if you delete an existing item in this knowledge set or you update something then you don't expect the results you get according to this new modification you've made in the data set.

3.2 Unsupervised Learning

It is quite different from supervised learning. Here we don't know about the labels (output) of different cases. And here, we train the model with patterns by finding similarities. And then these patterns become the cluster (see Figure 12-3) .

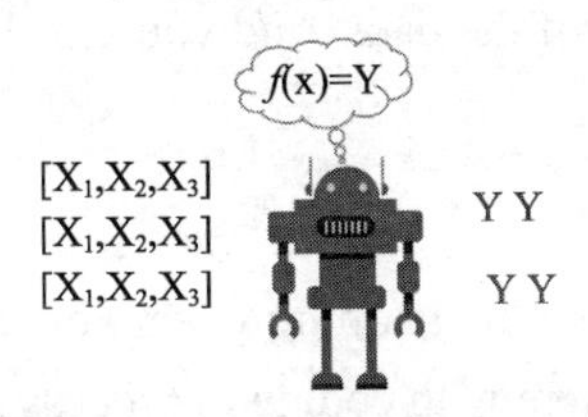

Figure 12-3 Unsupervised learning

Cluster = Collection of similar patterns of data

And then, this cluster is used to analyze and to process the data.

In unsupervised learning, we really don't know if the output is right or wrong. So here in this scenario, the system recognizes the pattern and tries to calculate the results until we get the nearly right value.

3.3 Reinforcement Learning

It is like reward-based learning. The example of reward based is, suppose your parents will give you a reward on the completion of a specific task. So here you know you've to complete this task and the time you need to complete it. The developer decides himself what reward he'll give on the completion of this task.

It is also feedback-oriented learning. Now you're doing some tasks and on the basis of these tasks, you're getting feedback. And if the feedback is positive then it means you're doing it right and you can improve your work on your own. And if the feedback is negative then you know as well what was wrong and how to do it correctly. And feedback comes from the environment where it is working.

It makes the system more optimal than the unsupervised scenario, because here we have some clues like rewards or good feedback to make our system more efficient.

4. Steps in Machine Learning

There are some key point steps of machine learning when we start to teach the machine.

4.1 Collect Data

As we already know machine learning is data oriented. We need data to teach our system for future predictions.

4.2 Prepare the Input Data

Now you've downloaded the data, but when you're feeding the data, you need to make sure of the particular order of the data to make it meaningful for you machine learning tool to process it, i.e. .csv file (comma separated value). This is the best format of the file to process the data because comma separated values help a lot in clustering.

4.3 Analyze Data

Now, you're looking at the patterns in the data to process it in a better way. You're checking the outliers (scope and boundaries) of the data. And you are also checking the novelty (specification) of the data.

4.4 Train Model

This is the main part of the machine learning when you are developing the algorithm where you are structuring the complete system with coding to process the input and give back the output.

4.5 Test the Model

Here you're checking the values you're getting from the system whether it matches your required outcome or not.

4.6 Deploy It in the Application

Let's discuss an example of autonomous cars which don't have human intervention, which run on their own. The first step is to collect the data, and you have to collect many kinds of data. You're driving the car which runs on its own. The car should know the road signs, it should have the knowledge of traffic signals and when people crossing the road, so that it can make the decision to stop or run in different situations. So we need a collection of images of these different situations, it is our collect data module.

Now we have to make the particular format of data (images) like csv file where we store the path of the file, the dimensions of the file. It makes our system processing efficient. This is what we called preparing the data.

And then it makes the patterns for different traffic signals (red, green, yellow), for different sign boards of traffic and for its environment in which car or people running around. Next it decides the outlier of these objects whether it is static (stopped) or dynamic (running state). Finally it can make the decision to stop or to move in the side of another object.

These decisions are obviously dependent upon training the model, and what code we write to develop our model. This is what we say training the model and then we test it and then we deploy it in our real world applications.

5. Applications of Machine Learning

Machine learning is widely used today in our applications.

- You might use the Snapchat or Instagram app where you can apply the different animal's body parts on your face like ears, nose, tongue etc. These different organs places at the exact right spot in the image, this is an application of machine learning.
- Google is a widely used AI, ML. Google Lens is an application. If you scan anything through Google lens then it can tell the properties and features of this specific thing.
- Google Maps is also using machine learning. For example, if you're watching any department store on the map, sometimes it is telling you how much something is priced, and how expensive it is.

New Words

action	[ˈækʃn]	*n.*行动，活动；功能，作用
train	[treɪn]	*v.*训练；教育；培养
situation	[ˌsɪtjʊˈeɪʃn]	*n.*（人的）情况；局面，形势，处境；位置
conclude	[kənˈklu:d]	*vt.*得出结论；推断出；决定
comprehensive	[ˌkɒmprɪˈhensɪv]	*adj.*广泛的；综合的
bulk	[bʌlk]	*n.*大块，大量；大多数，大部分
scenario	[səˈnɑ:rɪəʊ]	*n.*设想；可能发生的情况；剧情梗概
backend	[ˈbækend]	*n.*后端
random	[ˈrændəm]	*adj.*任意的；随机的 *n.*随意；偶然的行动
attract	[əˈtrækt]	*vt.*吸引；引起……的好感（或兴趣） *vi.*具有吸引力；引人注意
accurate	[ˈækjʊrət]	*adj.*精确的，准确的；正确无误的
validate	[ˈvælɪdeɪt]	*vt.*确认；证实
knowledge	[ˈnɒlɪdʒ]	*n.*知识；了解，理解
gradually	[ˈgrædʒʊəlɪ]	*adv.*逐步地，渐渐地
horizontally	[ˌhɒrɪˈzɒntəlɪ]	*adv.*水平地，横地
vertical	[ˈvɜ:tɪkl]	*adj.*垂直的，竖立的 *n.*垂直线，垂直面

approximately	[əˈprɒksɪmɪtlɪ]	*adv*.近似地，大约
reward	[rɪˈwɔ:d]	*n*.奖赏；报酬；赏金；酬金 *vt*.奖赏；酬谢
optimal	[ˈɒptɪməl]	*adj*.最佳的，最优的；最理想的
feed	[fi:d]	*vt*.馈送；向……提供
comma	[ˈkɒmə]	*n*.逗号
separated	[ˈsepəreɪtɪd]	*adj*.分隔的，分开的
novelty	[ˈnɒvltɪ]	*n*.新奇，新奇的事物
deploy	[dɪˈplɔɪ]	*v*.使展开；施展；有效地利用
intervention	[ˌɪntəˈvenʃn]	*n*.介入，干涉，干预
decide	[dɪˈsaɪd]	*vt*.决定；解决；裁决 *vi*.决定；下决心
static	[ˈstætɪk]	*adj*.静止的；不变的
organ	[ˈɔ:gən]	*n*.器官；元件

Phrases

as the name suggests	顾名思义
take a look into	看一看
be concerned with	涉及；与……有关
wireless communication	无线通信
knowledge set	知识集
straight line	直线
road sign	交通标志，路标
traffic signal	交通信号；红绿灯

Abbreviations

TB (TeraByte)	太字节

Text B 参考译文

什么是机器学习

人工智能和机器学习是当今最流行的技术。人工智能教导计算机表现得像人一样，可以思考并做出像人一样的反应，像人一样行事。

1. 什么是机器学习

顾名思义，机器学习意味着机器正在学习。

这是教会机器有关事物的技术。它是人工智能的一个分支，是人工智能的基础。在这里使用数据训练机器。如果仔细研究，就会发现它类似于数据挖掘。实际上，其背后的概念是机器学习和数据挖掘都是面向数据的。在两种情况下都处理数据。实际上，在数据科学或大数据中，我们分析数据并从中进行统计，然后研究如何维护数据，如何得出结果并对其进行总结，而不是维护完整的、全面的数据。但是在机器学习中，我们教会机器做出有关事物的决策。使用不同的数据集教机器，然后检查机器的某些情况，并查看从这种未知情况中得到的结果。还将这种训练好的模型用于新场景的预测。

我们用历史数据、观察结果和实验来教机器。然后，根据这些经验对机器进行预测并做出响应。

就像已经说过的那样，机器学习与数据挖掘和统计紧密相关。

- 数据挖掘：与数据分析有关。
- 统计：与预测/概率有关

2. 为什么需要机器学习

在这个时代，我们正在使用无线通信、互联网等。使用社交媒体、驾驶汽车或现在正在做的任何事情，实际上是在后端生成数据。如果对汽车如何生成数据感到惊讶，请记住每辆汽车内部都有一台小型计算机，可以完全控制车辆，即哪个组件何时需要电流、何时需要启动或切换特定组件。通过这种方式，可以生成 TB（太字节）的数据。

但是这些数据对于获得结果也很重要。举个例子，尝试清楚地理解这个概念。假设一个人住在城镇里，然后去购物商场买东西。一个产品有很多个。当他买东西时，现在可以生成他所买东西的模式。同样，可以生成不同人的销售和购买东西的模式。现在你可能正在考虑一个人随机来买东西，然后他再也不会来了，但是我们也有此类情况的模式。在这种模式的帮助下，可以决定人们最喜欢的事物以及他们何时再次来到购物中心。他们将在入口处看到他们想要的东西。这就是通过机器学习吸引客户的方式。

3. 机器如何学习

实际上，机器通过数据模式学习。让我们从数据的数据集开始。提供给机器的输入为 X，得到的响应为 Y。这里有三种学习类型。

- 监督学习。
- 无监督学习。
- 强化学习。

3.1 监督学习

在监督学习中，了解不同情况（输入）并且知道这些情况的标签（输出）。在这里，已经了解基本事实，因此仅关注函数（操作），因为它是这里最主要和最重要的事情（图12-1）。

（图略）

此时，仅创建函数以获取输入的输出。并且尝试创建处理数据的函数，并在大多数情况下尝试给出准确的输出（Y）（图12-2）。

（图略）

因为从输入的已知值开始，所以可以验证模型并将其改进。

现在，以不同的数据集教机器。此时，是在未知情况下进行检查并生成值。

注意：假设你为机器提供了某种数据集，现在可以根据该数据集训练模型。根据所提供的知识集从该模型获得结果。但是，如果你删除了该知识集中的现有项目或进行了更新，那就得不到结果，因为你修改了数据集。

3.2 无监督学习

它与监督学习完全不同。在这里，不了解不同情况的标签（输出）。此时，通过寻找相似性来训练带有模式的模型。然后这些模式成为聚类（图12-3）。

（图略）

聚类=收集相似模式的数据

然后，用该聚类分析和处理数据。

在无监督学习中，并不知道输出是否正确。因此在这种情况下，系统会识别出模式并尝试计算结果，直到获得接近正确的值为止。

3.3 强化学习

这就像基于奖励的学习。例如，假设你的父母会在你完成特定任务时给你奖励。因此，此时你知道必须完成此任务及所需的时间。开发人员可以在这里自行决定完成此任务将获得的奖励。

这也是面向反馈的学习。现在你正在执行一些任务，并在这些任务的基础上获得了反馈。如果反馈是正面的，则说明做对了，可以自己改善工作。如果反馈是负面的，那么你也知道问题出在哪里以及如何更正。反馈来自其工作环境。

它使系统比无人监督的方案更优化，因为这里有一些线索，例如奖励或良好的反馈可以使系统更高效。

4. 机器学习的步骤

当开始教机器时，有一些机器学习的关键步骤。

4.1 收集数据

众所周知，机器学习是面向数据的。我们需要数据来指导系统进行未来的预测。

4.2 准备输入数据

现在已经下载了数据，但是在输入数据时，需要确保数据的特定顺序，以使其对机器学习工具进行处理有意义，例如.csv 文件（逗号分隔值）。这是处理数据的最佳文件格式，因为逗号分隔的值在聚类中有很大帮助。

4.3 分析数据

现在正在查看数据中的模式，以便以更好的方式对其进行处理。检查数据的异常值（范围和边界）和数据的新颖性（规格）。

4.4 训练模型

开发算法时，这是机器学习的主要部分，在该算法中，将使用编码以构造完整的系统来处理输入并返回输出。

4.5 测试模型

在这里，要检查从系统获取的值是否与所需结果相匹配。

4.6 在应用程序中部署

让我们讨论一个自动驾驶的示例，该示例无须人工干预，而是自行运行。第一步是收集数据，必须收集多种数据。驾驶的车自主运行。该汽车应该知道路标，应该了解交通信号以及人们过马路的知识，以便可以在不同情况下做出停车或行驶的决定。因此，需要收集这些不同情况的图像，这就是收集数据模块。

现在，必须制作特定格式的数据（图像），例如.csv 文件，在其中存储文件的路径及文件的大小。它使系统处理效率更高。这就是准备数据。

然后，它为不同的交通信号（红色、绿色、黄色），不同的交通标志牌以及汽车或人员在附近奔波的环境制定模式。其次，确定这些对象的异常值是静态的（已停止）还是动态的（运行状态）。最后，它可以决定停止还是移到另一个对象的侧面。

这些决定显然取决于训练模型以及为开发模型编写的代码。这就是所说的训练模型及测试模型，然后将其部署到实际应用程序中。

5. 机器学习的应用

如今，机器学习已在应用程序中广泛使用。

- 可以使用 Snapchat 或 Instagram 应用程序把动物的不同身体部分（例如耳朵、鼻子、舌头等）放到你的面部图像上。将这些不同的器官放在图像的正确位置，就是机器学习的一种应用。

- Google 广泛使用人工智能和机器学习。Google Lens 是一个应用程序。如果通过 Google 镜头扫描任何东西，它就能告诉此特定事物的属性和功能。
- Google Maps 也正在使用机器学习。例如，如果你正在看地图上的任何百货商店，有时它会告诉你价格多少，有多贵。

Exercises

[Ex. 1] Answer the following questions according to Text A.

1. What is artificial intelligence (AI)? What do these processes include?
2. What do particular applications of AI include?
3. What can AI be categorized as? What are they?
4. What does AI as a service allow individuals and companies to do?
5. What do researchers and marketers hope the label augmented intelligence will do?
6. How many types does Arend Hintze categorize AI into? What are they?
7. What are the different types of technology AI is incorporated into?
8. How many types of machine learning algorithms are there? What are they?
9. What are the areas AI has made its way into?
10. In 2016,what did the National Science and Technology Council (NSTC) do?

[Ex. 2] Fill in the following blanks according to Text B.

1. Machine learning means ____________________. It is ______________ through which we teach the machines about things. It is a branch of __________________.

2. We teach the machine with ________________, ______________ and ________________. And then, we predict with the machine from these learnings and ________________.

3. Machine learning is closely related to ________________ and ________________. Machine learning is ______________.

4. In supervised learning, we know about ________________ and we know ________________.

5. Because we've started with known values for our inputs, we can ________________ and ________________.

6. In unsupervised learning, we really don't know ______________________.

7. Reinforcement learning is like ____________________. It is also ______________________.

8. There are some key point steps of machine learning when we start to teach the machine. They are ____________, ____________, ____________, ____________, ____________, and ______________.

9. ________________________ is the best format of the file to process the data. Because

comma separated values help a lot in ____________.

10. Train model is the main part of the machine learning when you are ______________ where you are structuring the complete system with coding to____________ and ______________.

[Ex. 3] Translate the following terms or phrases from English into Chinese and vice versa.

1. deep learning algorithm	1.
2. confidence scoring schema	2.
3. expert system	3.
4. image recognition	4.
5. reinforcement learning	5.
6. *n.*聊天机器人	6.
7. *n.*获得	7.
8. *adj.*认知的，认识的	8.
9. *n.*检查，检测	9.
10. *vt.*诊断；判断 vi.做出诊断	10.

[Ex. 4] Translate the following sentences into Chinese.

Strong Artificial Intelligence (Strong AI)

Strong artificial intelligence (strong AI) is an artificial intelligence construct that has mental capabilities and functions that mimic the human brain. In the philosophy of strong AI, there is no essential difference between the piece of software, which is the AI, exactly emulating the actions of the human brain, and actions of a human being, including its power of understanding and even its consciousness.

Strong artificial intelligence is also known as full AI.

Strong artificial intelligence is more of a philosophy rather than an actual approach to creating AI. It is a different perception of AI wherein it equates AI to humans. It stipulates that a computer can be programmed to actually be a human mind, to be intelligent in every sense of the word, to have perception, beliefs and have other cognitive states that are normally only ascribed to humans.

However, since humans cannot even properly define what intelligence is, it is very difficult to give a clear criterion as to what would count as a success in the development of strong artificial intelligence. Weak AI, on the other hand, is very achievable because of how it stipulates what intelligence is. Rather than try to fully emulate a human mind, weak AI focuses on developing intelligence concerned with a particular task or field of study. That is a set of activities that can be broken down into smaller processes and therefore can be achieved in the scale that is set for it.

[Ex. 5] Fill in the blanks with the words given below.

interact	compare	doctors	search	pilot's
finance	choices	applications	respond	make

What Are the Applications of Artificial Intelligence?

There are many different applications of artificial intelligence and new uses for the technology are developed each year as artificial intelligence programs become more sophisticated. Artificial intelligence is frequently used by the military and in aviation and robotics. It is also used in the public sector in medicine, ___1___, and business. Artificial intelligence has even made intelligent toys available for children at relatively low prices.

One of the most obvious ___2___ of artificial intelligence is in creating computerized brains for robots. These programs allow robots to make choices about how to respond to stimuli, without direct input from humans. One example of this type of technology is the rovers that are deployed on the surfaces of distant planets. These machines make ___3___ about how to get around obstacles because the great distance between the robot and humans makes it impractical for the robot to wait for each new instruction to arrive. Less sophisticated versions of these programs are often used in toys, creating robots with a limited ability to ___4___ with their environment and with the children playing with them.

There are also applications of artificial intelligence in the field of medicine. Intelligent diagnostic programs can make observations about a patient's symptoms and ___5___ that data to possible syndromes that the patient could be affected with. This technology is extremely useful because it is easy for human ___6___ to overlook a condition, whereas computer programs cannot forget to take information into account.

Military organizations also find many applications for artificial intelligence. Programs are used to run simulations which can help humans ___7___ important tactical decisions. Additionally, artificial intelligence is used in aviation, where it can assist in training pilots. These programs can also gather information about a ___8___ abilities, which helps it learn to train each pilot more efficiently.

Many people come into contact with artificial intelligence in customer service programs. These programs, often called chat bots, are programmed to ___9___ to inquiries from clients and offer support or answers to questions. Human customer service representatives are often made available if the artificial intelligence program is unable to help.

Financial institutions also find applications for artificial intelligence. Specialized programs are designed to ___10___ for patterns in financial markets in order to make investment decisions. These programs are capable of initiating trades on their own and often make decisions that are at least as financially sound as those of human brokers.

Online Resources

二维码	内　容
	计算机专业常用语法（12）：同位语和插入语
	在线阅读（1）：Deep Learning （深度学习）
	在线阅读（2）：VR vs AR （虚拟现实对增强现实）

图书资源支持

感谢您一直以来对清华版图书的支持和爱护。为了配合本书的使用，本书提供配套的资源，有需求的读者请扫描下方的“书圈”微信公众号二维码，在图书专区下载，也可以拨打电话或发送电子邮件咨询。

如果您在使用本书的过程中遇到了什么问题，或者有相关图书出版计划，也请您发邮件告诉我们，以便我们更好地为您服务。

我们的联系方式：

地　　址：北京市海淀区双清路学研大厦 A 座 714

邮　　编：100084

电　　话：010-83470236　010-83470237

客服邮箱：2301891038@qq.com

QQ：2301891038（请写明您的单位和姓名）

资源下载：关注公众号“书圈”下载配套资源。

资源下载、样书申请

书圈

获取最新书目

观看课程直播